# CANADA'S WASTE FLOWS

# CANADA'S WASTE FLOWS

MYRA J. HIRD

McGill-Queen's University Press
Montreal & Kingston • London • Chicago

© McGill-Queen's University Press 2021

ISBN 978-0-2280-0528-5 (cloth)
ISBN 978-0-2280-0645-9 (ePDF)
ISBN 978-0-2280-0646-6 (ePUB)

Legal deposit first quarter 2021
Bibliothèque nationale du Québec

Printed in Canada on acid-free paper that is 100% ancient forest free (100% post-consumer recycled), processed chlorine free

We acknowledge the support of the Canada Council for the Arts.
Nous remercions le Conseil des arts du Canada de son soutien.

---

Library and Archives Canada Cataloguing in Publication

Title: Canada's waste flows / Myra J. Hird.
Names: Hird, Myra J., author.
Description: Includes bibliographical references and index.
Identifiers: Canadiana (print) 2020035373X | Canadiana (ebook) 20200353810 | ISBN 9780228005285 (cloth) | ISBN 9780228006459 (ePDF) | ISBN 9780228006466 (ePUB)
Subjects: LCSH: Refuse and refuse disposal—Canada. | LCSH: Recycling (Waste, etc.)—Canada.
Classification: LCC TD789.C3 H57 2021 | DDC 363.72/80971—dc23

---

This book was typeset in 10.5/13 Sabon.

*I dedicate, as always, this book to Inis and Eshe. This pilot light is there in your pockets. To my parents, who taught me the most about consuming the least. And to Christophe, for the path that we share, and the lives we don't want to waste.*

# Contents

Table and Figures  ix

Acknowledgments  xi

SECTION ONE: WASTE IN SOUTHERN CANADA

1 Canada's Waste Flows  3

2 Southern Canada's Waste Management Problem  33

3 Looking for Redemption in All the Wrong Places: Hiding Our Waste and the Cult of Recycling  58

SECTION TWO: CANADA'S ARCTIC WASTES

4 Canadian Settler Colonial Waste  87

5 Arctic Wasteland  111

6 Wasting Animals  126

SECTION THREE: WASTE AT THE END OF THE WORLD

7 A Good Soup  149

8 Wasting (in) the Anthropocene  166

9 The Indeterminate Material Politics of Waste  186

Epilogue: Canada's Waste Future  215

Notes  233

References  245

Index  307

# Table and Figures

### TABLE

3.1 Emissions from heavy-duty diesel vehicles (HDDVs)  80

### FIGURES

1.1 November 2017 protest in Manila, Philippines. Photographer unknown, courtesy of EcoWaste Coalition. http://ecowastecoalition.blogspot.com.  4

2.1 Recycling collected at the Kingston Area Recycling Centre. Photo by the author.  46

3.1 The waste hierarchy. Graphic adapted from Recycle More. https://www.recycle-more.co.uk/business-zone/office-recycling-and-waste-management/the-waste-hierarchy.  60

3.2 Garbage can, park in Kingston, Ontario. Photo by the author.  62

3.3 The contribution of recycling to the supply of metals and minerals. Reproduced from Luis Tercero Espinoza, "Criticality of Mineral Raw Materials and Sustainability Assessment," 2012. https://eplca.jrc.ec.europa.eu/uploads/rawmat-Tercero-criticality-of-mineral-raw-materials-and-sustainability-assessment.pdf.  66

4.1 Anti-littering poster, Pangnirtung, Nunavut.
Photo by the author. 88

5.1 Abandoned military installation and solid waste disposal sites near Iqaluit. Map reproduced from *Environmental Study of a Military Installation and Six Waste Disposal Sites at Iqaluit*, NWT, Volume 1: *Site Analysis*, Volume 2: *Appendices* (Victoria, BC: Royal Roads Military College, 1995). Reproduced in Terry Dobbin, "Iqaluit Needs Help with Contaminated Clean-Ups," *Nunatsiaq News*, 14 January 2013. 113

5.2 Garbage at Cape Dyer, Nunavut DEW Line site. Source unknown, photo courtesy Margo Pfeiffer, "Fixing the Mess," *Up Here* (October/November 2012): 52–9. 120

5.3 "Goody bags" of PCB-laced soil await shipment to southern incinerators from Cape Dyer DEW site.
Photo courtesy Margo Pfeiffer. 121

6.1 Hawk used at a landfill, servicing Montreal, Quebec.
Photo by the author. 133

6.2 *Qimmiit* and a raven on the Iqaluit dog lot moved to make way for the expanding Iqaluit airport. Photo by the author. 141

9.1 Julie Alivaktuk in front of the West 40 dump in Iqaluit, Nunavut. Photo by Alivaktuk's father in law. 188

E.1 Trash Pack toys. Photo by the author. 220

# Acknowledgments

This book has been my companion for several years, and family, friends, and colleagues have immeasurably helped me see it to fruition. I thank the people living in Iqaluit and Pangnirtung. I very much thank my patient editor, Jonathan Crago at McGill-Queen's University Press. I thank Kim Renders for her enthusiastic faith in my work and life. I thank, and am grateful to, the graduate students who have journeyed with me: Alexander Zahara, Scott Lougheed, Jessica Metuzals, Cassandra Kuyvenhoven, Diana Van Vlymen, Hillary Predko, Micky Renders, and Jacob Riha. I thank my friends and colleagues near and far: Judy Haschenburger, Allison Rutter, Anne McRory, Hillary Warder, Claire Colebrook, Lisa Smyth, Lynda Williams, Wendy Craig, Mary Ellen Kenny, Ilse Borgers, Liz Frazer, Cecilia Åsberg, Tora Holmberg, Kate Scantebury, Jacob Bull, Charlotta Råsmark, Anita Hussénius, Ann-Sofie Lönngren, Annica Gullberg, Nils Johansson, Björn Wallsten, Donna Haraway, Jamie VanGulck, Karen Barad, Anna Tsing, Kathryn Yusoff, Nigel Clark, Peter van Wyck, Sabrina Perić, Romain Garcier, Laurence Rocher, Nathalie Ortar, Michel Lussault, Delphine Vernet, Michael Civet, Morgan Rossignol, Margorie Bourgeonnier, Phil Selliez, Maddi Frances, Marielle Garriga, Karine Monier, Lisa Bishop-Stanton, Eve Stanton, Mireille Villain, Pierre Founaud, Sophie Coudrain, Roland deJong, Anette and Jean-Claude Merle, Jane Vergniaud, Michel Tordjman, Mike Uzan, Kirk Fontaine, Elaine Power, and the Sinclairs.

Earlier versions of part of chapter 1 first appeared in two articles: Scott Lougheed, Jessica Metuzals, and Myra J. Hird, "Modes of Governing Canadian Waste Management: A Case Study of Metro Vancouver's Energy-from-Waste Controversy," *Journal of*

*Environmental Policy and Planning* 20, no. 2 (2017), 1–14; and Myra J. Hird, "Knowing Waste: Toward an Inhuman Epistemology," *Social Epistemology* 26, nos. 3–4 (2012), 453–69. Chapter 2 is adapted from an earlier version of Myra J. Hird, Scott Lougheed, R. Kerry Rowe, and Cassandra Kuyvenhoven, "Making Waste Management Public (or Falling Back to Sleep)," *Social Studies of Science* 44, no. 3 (2014), 441–65. Chapter 3 is adapted from an earlier version of Myra J. Hird and Cassandra Kuyvenhoven, "Waste Transportation and the Downstream-Recycle, Upstream-Reduce Equation: Waste's Management within Neoliberal Capitalism," in *Opening the Bin*, edited by Richard Ek (Cambridge, UK: Cambridge Scholars Press, 2020), 79–97. Chapter 4 is adapted from an earlier version of Myra J. Hird and Alexander Zahara, "The Arctic Wastes," in *Anthropocene Feminism*, edited by Richard Grusin (Minneapolis, MN: University of Minnesota Press, 2017), 121–45. Chapter 5 is adapted from an earlier version of Myra J. Hird, "Waste Legacies: Land, Waste, and Canada's DEW Line," *Northern Research* 42 (2016), 173–95. Chapter 6 is adapted from an earlier version of Alexander Zahara and Myra J. Hird, "Raven, Dog, Human: Inhuman Colonialism and Unsettling Cosmologies," in the "Learning How to Inherit in Colonized and Ecologically Challenged Lifeworlds" special issue of *Environmental Humanities* 7 (2016), 169–90. Chapter 7 is adapted from an earlier version of Myra J. Hird, "Burial and Resurrection in the Anthropocene: Infrastructures of Waste," in *Infrastructures and Social Complexity: A Routledge Companion*, edited by Penelope Harvey, Casper Jensen, and Atsuro Morita (London, UK: Routledge, 2017), 242–52; and Myra J. Hird, "The Phenomenon of Waste World Making," in *Rhizomes: Cultural Studies in Emerging Knowledge* 30 (2016), 1–8. Chapter 8 is adapted from an earlier version of Nigel Clark and Myra J. Hird, "Deep Shit," *O-Zone: A Journal of Object-Oriented Studies*, "Objects/Ecology" special issue, edited by Eileen Joy and Levi Bryant, 1 (2014), 44–52; and Myra J. Hird and Kathryn Yusoff, "Subtending Relations: Bacteria, Geology, and the Possible in the Anthropocene," in *Genealogies of Speculation*, edited by Suhail Malik and Armen Ananessian (London, UK: Bloomsbury Press, 2015), 319–42.

The manuscript also draws on previous publications of my own, including Myra J. Hird, "Waste, Environmental Politics and Dis/Engaged Publics," *Theory, Culture and Society*, "Geo-Social Formations" special issue, edited by Nigel Clark and Kathryn

Yusoff, 34, nos. 2–3 (2017), 187–209; Myra J. Hird, "Waste, Landfills, and an Environmental Ethic of Vulnerability," *Ethics and the Environment* 18, no. 1 (2013), 105–24; and Myra J. Hird, "In/human Waste Environments," GLQ: *A Journal of Lesbian and Gay Studies*, special issue on Queer Inhumanisms, 21, nos. 2–3 (2015), 213–15. All text is reproduced with permission.

# SECTION ONE

# Waste in Southern Canada

CHAPTER ONE

## Canada's Waste Flows

> Garbage is all that anonymous stuff falling between valued objects and simple dust.
>
> Kennedy 2007, 7

Let's begin with a few snapshots of Canada's waste issues:

### SNAPSHOT A

In early February 2013, sixty-nine shipping containers of waste sourced from Vancouver, Canada, arrived in Manila, Philippines, via Chronic Incorporated, a plastics exporter based in Whitby, Ontario (Ballingall 2014). The containers had been officially declared scrap materials for recycling. However, customs officials found that the 2,500 tonnes[1] of recycling was in fact "household garbage, soggy paper, and even used adult diapers" – a heterogeneous mix of what the waste management industry calls municipal solid waste (MSW) (Hopper 2015, 6). After three years of decomposing in the tropical Philippines weather, the shipment began leaking "garbage juice" around the crowded containment terminal (Hopper 2014, 2). In July 2015, twenty-nine of the containers were transported to a landfill in the province of Tarlac. The Philippines government spent upwards of C$2 million to temporarily store the rest of this particular shipment of Canada's exported waste (Hopper 2014).

In late 2014, the Philippines government recommended that the containers return to Canada under the provisions of the Basel Convention ([1989] 2011), which prohibits developed countries from shipping waste to developing nations. Environment and Climate Change

Figure 1.1 | Protest in Manila, Philippines, November 2017.

Canada, the federal government agency that oversees the movement of waste across international borders, countered with the argument that under the Basel Convention only toxic waste transportation requires a specific license (Prakash 2015). Moreover, Canada had no domestic laws that the federal government might invoke to compel Chronic Incorporated to return the containers to Canada (Ministry of Environment 2015). Canada's Department of Foreign Affairs reassuringly claimed that they had worked with the Philippines government to find a solution, and that the local government there "considered options to dispose of the shipment in accordance with its laws and regulations" (ibid., 11). But frustrated by the inertia of the political and legal back-and-forth (if not the waste itself), Leah Paquiz, a Philippines member of the House of Representatives, made a succinct press statement: "Pick up your garbage Canada, and show us the decency that we so rightly deserve as a nation. My motherland is not a garbage bin for Canada" (Hopper 2014, 3).

Intending to move Canada out of its torpor, in September 2016, Manila Regional Trial Court branch judge Tita Bughao Alisuag ordered Chronic Incorporated to ship the waste back to Canada at the waste exporter's own expense (CBC Radio *Current* 2019b). But Jim Makris, owner of Chronic Incorporated, claimed that the contents he purchased from a recycling firm in Vancouver were 95 per cent plastic, with a small amount of paper and aluminum that would be found in a standard household recycling bin (Goodman 2015). Makris denied that the sixty-nine containers included household waste and maintained that he did not ship the waste overseas, as, he argued, "it costs $40 per tonne to dump garbage in Canada, but $80 per tonne to ship his recycling material to the Philippines" (Ballingall 2014).[2] The global traffic of waste, it appears, has oscillating degrees of surplus value, it is hard to track, and evidently it is difficult to apportion responsibility for its management.

The Canadian shipment received widespread media attention and has become emblematic of Canada's worsening problem with solid waste management and disposal (CBC Radio *Current* 2019a). In response to the incident, Canadian prime minister Justin Trudeau issued the following statement:

> Going forward we need to ensure that if a situation like this were to arise again, that the Canadian government had more power to demand action from the companies responsible. I believe there are loopholes here that were allowed to be skirted that we need to make sure we close both for Canada's interest and good relationships with our neighbors. (In Garcia 2015, 6)

On 29 June 2019, sixty-nine containers arrived in Vancouver, BC, scheduled to be shipped to the energy-from-waste (EfW) facility in Burnaby (Boynton 2019). Vancouver's waste travelled roughly 20,072 kilometres on trucks and ships using non-renewable fossil fuels, around the globe and back again, to be disposed of at a cost of some C$1.14 million.

SNAPSHOT B

The Giant Mine, located on the Ingraham Trail, close to Yellowknife in the Northwest Territories, is the most well-known abandoned mine in Canada. Claims to the Giant Mine were staked back in 1936, and the mine was brought into full production from 1948 to 2004, producing over seven million ounces of gold. Mining gold at this site consisted of roasting arsenopyrite ore, which produces arsenic trioxide dust – a highly toxic form of arsenic – as a by-product. In 1999, Royal Oak Mines, which had owned the mine since 1990, filed for bankruptcy, and the mine rights were sold to Miramar Mining, which, with its environmental liabilities severed, then used the Reclamation Security Agreement to terminate its obligations when the mine closed five years later. After the mine was finally abandoned in 2005, it became the responsibility of the Ministry of Indian Affairs and Northern Development – that is, it officially became a public liability for which remediation is now estimated to cost C$1 billion (CBC News 2013b). Besides the approximately one hundred on-site buildings (many of which contain hazardous materials), the eight open pits, and the contaminated soils and waste rock around the mine, there are some 237,000 tons of arsenic trioxide dust to remediate.

After considering some fifty-six methods of "take it out" versus "leave it in" technologies, industry and government decided on the frozen-block method (FBM). This is by no means an immediate solution, as it will take about twenty years to complete in various stages. FBM is being used in several mines in Canada, whereby mining waste (rich in arsenic and other dangerous elements, minerals, and metals) is frozen in perpetuity, with the hope that a future generation will safely resolve this ongoing waste issue. The "in perpetuity" time frame is alarming local residents, especially as the mine is on Yellowknives Dené territory, and the lengthy stages of the environmental assessment are by no means concluded (to date I count some 664 documents relating to the Giant Mine waste remediation assessment). The board reviewing the plans noted public concern over the project's indefinite time frame: "Given enough time, the project is eventually likely to cause significant adverse effects" (Livingstone 2013). In a similar vein, a petition from Alternatives North asks of the Office of the Auditor General:

> Please provide a detailed justification for the trade-offs that were made in choosing the frozen block method for arsenic containment at the Giant Mine *even though it requires perpetual care forever and how the needs of future generations were considered.* (O'Reily 2013, 2; my emphasis)

Protestors of FBM argue that it amounts to nothing more than a "holding pattern" – like nuclear waste – because it projects a final solution onto a future generation. "In perpetuity" cannot factor in the effects of, for instance, climate change or biodiversity loss. It cannot predict, prevent, or remediate ecological disasters resulting from massive flooding, permafrost melting, or earthquakes. In the here and now, most mining infrastructure – buildings and built structures, transportation networks, slope stability, tailings and water-retention structures, and site hydrology – have been designed assuming no change to the climate, and only limited adoption of climate change plans are underway (Pearce et al. 2011, 348, 363).[3] Climate change is low on the list of priorities for mining operators because mines have a relatively short life span and will not be operating when the more severe effects of climate change manifest themselves in the future. Planning for post-operational and closure stages also requires a significant economic, organizational, and political commitment to the future – one that mining companies are not keen to engage in because of the cost. Nor does "in perpetuity" take political affiliations and priorities of the future into account: will the Canadian government (assuming there still is one) prioritize FBM to the tune of some C$1.9 million annually, in perpetuity? Or might there be other environmental issues pressing upon the government's attention?

Hedging their bets, the mining corporation BachTech embedded the following within their press release announcing their contract to clean up arsenopyrite at Northern Manitoba's Snow Lake:

> Certain statements included or incorporated by reference in this news release, including information as to the future financial or operating performance of the Company, its subsidiaries and its projects, constitute forward-looking statements. The words "believe," "expect," "anticipate," "contemplate," "target," "plan," "intends," "continue," "budget," "estimate," "may," "schedule" and similar expressions identify forward-looking statements. Forward-looking statements include, among other things, statements regarding targets, estimates and assumptions in respect of gold production and prices, operating costs, results and capital expenditures, mineral reserves and mineral resources and anticipated grades and recovery rates. Forward-looking statements are necessarily based upon a number of estimates and assumptions that, while considered reasonable by the Company, are *inherently subject to significant business, economic, competitive, political and social uncertainties and*

*contingencies*. Many factors could cause the Company's actual results to differ materially from those expressed or implied in any forward-looking statements made by, or on behalf of, the Company. *Investors are cautioned that forward-looking statements are not guarantees of future performance and, accordingly, investors are cautioned not to put undue reliance on forward-looking statements due to the inherent uncertainty therein*. Forward-looking statements are made as of the date of this press release and the Company *disclaims any intent or obligation* to update publicly such forward-looking statements, whether as a result of new information, future events or results or otherwise. (2013, 1; my emphasis)

While this disclaimer acknowledges the speculative nature of forward-looking statements with regard to profits, little is said about the significant material uncertainties here. These uncertainties relate not only to the amount and severity of waste (contaminated or not) produced, but to interacting geological thresholds. BachTech's disclaimer inadvertently calls our attention to the inability to completely identify and define the pre-given processes that are not only affecting contemporary waste practices, but will affect future ones as well.

SNAPSHOT C

Metro Vancouver (MV) consists of twenty-one municipalities and manages over 1.3 million tons of MSW annually (Metro Vancouver 2013). MV's current waste disposal system includes the Vancouver Landfill operated by the City of Vancouver, the Cache Creek Landfill operated by the private company Belkorp, and an EfW facility owned by MV and operated by Covanta (another private company). MV is responsible for providing waste disposal services for the municipalities under its remit. MV generates revenue from its own EfW facility and transfer stations through tipping fees, which are a cost-per-tonne paid to the operator of the facility to dispose of waste there. However, MV also relies heavily on landfills, transfer stations, and transport provided by other companies, the cost of which far outweighs the revenue MV earns from its own infrastructure. For example, in 2014, solid waste tipping fees generated C$89.2 million in revenue for MV, yet in the same year MV spent over C$100 million on MSW management overall (Metro Vancouver 2014b). Private waste haulers exporting waste out of the region (because it increases profit for the companies), a rapidly growing trend, has contributed to this revenue shortfall.

Like other Canadian municipalities, MV responded to dwindling landfill space and increasing costs by initiating an ambitious waste diversion campaign and proposing an expansion of EfW capacity concurrent with this diversion campaign (Metro Vancouver 2014a; Metro Vancouver 2010). Significant opposition to previous proposals to expand the region's two landfills and the proposal to increase the region's EfW capacity put the proposals on hold. Thus, the future of waste management in the MV region is both uncertain and embedded in a complex set of conflicting stakeholder interests.

Since 1966, the City of Vancouver has owned and operated the Vancouver Landfill, with additional waste sent to the Cache Creek Landfill, controlled by the village of Cache Creek (population 1,040) and Wastech Services (a Belkorp subsidiary), located 340 kilometres east of MV (Metro Vancouver 2007). The Cache Creek Landfill is contracted to serve MV, which reserves the right to deny any other municipality access to the landfill. The Village of Cache Creek receives royalties based on the total tonnage disposed, which has had a profound positive economic impact on the community. To wit, Cache Creek's long-term debt has been eliminated, overall revenues have increased for local government, and property taxes have decreased (Urban Systems 2002). In 2010, the BC government approved a pair of expansions to the Cache Creek Landfill proposed by the village and the landfill operator, Wastech, extending its life as a source of revenue for the town of Cache Creek by up to twenty-five years and making the landfill accessible to all municipalities (Sinoski 2010). Meanwhile, as part of a broader strategy to localize waste processing, MV proposed an aggressive goal to increase their EfW capacity by up to 500,000 tons by 2015, effectively replacing their reliance on the Cache Creek Landfill and on exporting waste to Washington State (Metro Vancouver 2010). From its inception, the proposal was divisive among different levels of government, local governments and residents in adjacent regions, and industry. In December of 2015, MV announced that it was halting plans to build a new EfW facility, citing concerns about uncertain future waste volumes due to improved waste reduction and recycling efforts (Nagel 2015).

## INTRODUCTION

As these brief snapshots suggest, Canada has a waste problem. The fact that some Canadian household waste turned up in the Philippines, Cambodia, and Malaysia has (temporarily at least) laid bare the complex material, economic, and political systems preoccupied with waste. Canada, like other countries, likes to move its waste around. We move our waste from region to region within Canada, and around the globe. Canada's waste flows. The Giant Mine controversy in the Northwest Territories calls our attention to a number of profound issues, not least of which is the temporality of waste. While economic projections focus on the financial dividends of all of the metals (gold, copper, etc.), oil, and gas that we are extracting, the Giant Mine illustrates that extractive industries are creating vast quantities of often highly toxic waste that require intergenerational storage, if not remediation. The longevity of mining waste (and certainly other types of waste, such as radioactive waste) means that waste issues intersect with current politics, economics, and sociocultural considerations, but also into the distant future. For instance, the Canadian federal government has agreed to foot the annual bill for the Giant Mine's frozen storage of arsenic trioxide waste – but are we confident that future governments will not have other pressing economic considerations due to, for instance, things like peak oil and climate change? And MV's ongoing debates about what to do with their waste echoes in communities – big and small – across Canada. All kinds of waste is piling up, and municipal governments are well aware that both waste volumes and the different types of (increasingly toxic) waste are increasing faster than diversion (which overwhelmingly consists of recycling) is able to mitigate. The most up-to-date engineered landfills have lifespans of up to thirty years, after which the probability of containment becomes indeterminate. And all nuclear waste in Canada is, by definition, stored temporarily.

We Canadians produce an enormous amount of waste. In 2009, and again in 2013, the Conference Board of Canada published an alarming statistic: Canada is the *world's leading producer* of MSW. Between 1990 and 2005, Canada's per person MSW production increased by 24 per cent. By 2000, Canadians were producing more waste per person than Americans, and by 2005, Canada was generating 791 kilograms of MSW per person – well above the OECD seventeen-country average of 610 kilograms per person, and almost twice as much as Japan (CBC News 2013a). This translates into 35 million tons of waste in a single calendar year. The latest available statistics report that Canadians produce annually an average of 777 kilograms of waste per capita (Statistics Canada 2012a). Even if Canada's overall recycling rate of 30 per cent is considered, we are still producing more than Japan's overall gross waste production (Ghoreishi 2013, 8). With the inclusion of the waste from the industrial, commercial, and institutional (ICI) sectors – including mining and agriculture – Canada produces nearly 1.3 billion tons of waste annually (ibid.).

WHAT WASTE IS

Organizations and government departments such as Statistics Canada and the Canadian Council of Ministers of the Environment define several types of waste and gather statistical information concerning production and diversion, and so on, including MSW; ICI waste, which may include medical wastes; construction, renovation, and demolition (CRD) waste; household hazardous and special wastes (HHSW); organics; packaging and printed paper (PPP), including paper and plastics; agricultural wastes, including things like animal sewage lagoons; extraction wastes, including mining (sometimes the Alberta oil sands gets its own category); and nuclear wastes.

Behind these apparently rather neat categories lies a complex set of assumptions, values, and compromises. As this book will detail, how we define waste has a lot to do with the politics and economics of consumption, intergovernmental and industry-government relations, urban-rural divides, health, labour relations, gender and waste economies, science-public relations, risk, governance, and so on (Maclaren and Nguyen 2003). Indeed, "waste exist[s] in the twilight zone where no clear, 'natural' definition of [it] can be given, within wide margins of uncertainty and variation" (Wynne 1987, 1). This is true in both a cultural-symbolic and a material sense. Waste is an inherently ambiguous linguistic signifier: anything and everything can become waste, and things can simultaneously be and not be waste, depending on the perceiver.[4] Or as Richard Wilk puts it, "If we make a simple list of all of the kinds of things that can be called waste, from an idle moment to a mountainous landscape, it should be clear that we are dealing with a folk-concept rather than an objective analytic category" (2015, 234).

Waste, in other words, is a generous linguistic, symbolic, and material signifier. We waste things both tangible and intangible, including love, time, youth, and energy. Supermodels, actresses, and drug addicts "waste away." Within capitalist societies, waste is often associated with resources that are out of place, and which span the gamut of lost work time and incomplete production to the failure to maximize profit through the non-use of potentially usable objects such as land. As Catherine Alexander and Andrew Sanchez point out, "surplus is ... integral to the capitalist process, creating and maintaining profit, and wasting human lives" (2019, 8). By contrast, Alexander Judd, former executive vice president of the American Can Company, defines waste as any area of land not being used for landfilling:

> The public perceives that the garbage crisis is caused by the runaway growth of disposables, packaging, and discards in general. The real problem, of course, is not the growth of garbage or the quantity of garbage; it is the closing of landfills and the failure to provide replacement sites or alternate ways to handle the discards of towns and cities. The production of garbage responds to growth in population, household formations, affluence, and commercial activity, but the capacity for the disposal of waste depends more on the availability of land – space – than any other factor (2005, 21).

Thus, for Judd, the problem is not that we discard increasing volumes of objects or that these objects are stashed underground: "wastefulness" is not using available land to indefinitely store waste.

We encounter an increasing number of objects as only a "fleeting presence" (Kennedy 2007, xi). Single-serving objects such as coffee cups, sugar packets, stir sticks and so on are determinate: we use them once, toss them in the garbage, and forget about them. We even have single-serving friends, according to Edward Norton's character in the movie *Fight Club*. Our disposable society is, as Jean Baudrillard argues, the "gloomy, bureaucratic caricature in our societies, where wasteful consumption has become a daily obligation, a forced and often unconscious institution like indirect taxation, a cool participation in the constraints of the economic order" (1998, 43). We are, in short, a disposable society. Indeed, within our contemporary "throwaway society ... we encounter objects as essentially evanescent, instantaneous" (Kennedy 2007, 158). Yet while most disposables appear the same before and after their use – my take-out plastic coffee lid looks pretty much the same when I buy it and when I toss it – their ontology has fundamentally changed. Before use, the object is a desirable commodity; afterwards it's garbage (ibid.). As such, no entity is in its essence waste, yet everything is potentially waste.

In "Combing through Trash: Philosophy Goes Rummaging," Elizabeth Spelman points out that despite our relative silence on the subject, waste nevertheless says a great deal about who we are (2011, 313–25). Firstly, making waste is something we do every day yet rarely talk about. Indeed, waste is a testimonial to a desire to forget. Diligent individual practices of placing waste on sidewalks to be taken to dumping stations, landfills, incinerators, and the like ritualizes this forgetting – a forgetting that is made possible by legislative decision, regulative enforcement, risk models, community accession, and engineering practice. As such, waste repositories make their appearance on and in the landscape as a material enactment of forgetting. Of course, we have to know about something in order to forget it. In other words, as Joshua Reno (2016) observes, we need to know what waste we are producing before we can figure out what to do with it. And here lies one of the most challenging aspects of Canada's waste. The types of waste with which Canadians are most familiar – the waste that we encounter directly, if only momentarily, in the act of sorting it for curbside pickup – represents very little of our total waste production. The overwhelming majority of waste in Canada, produced by industry through processes of extraction and production, remains completely beneath our radar. This is less a case of forgetting, then, than being completely unaware of the nature of our waste problem.

"[A] society preoccupied with concealing its wastes," Kennedy observes, "must have something important to hide from itself" (2007, 4). The desire to disgorge ourselves of waste and remove it from sight has psycho-social, evolutionary, cultural, and structural explanations. Mary Douglas (2007) provided a well-known theory of community development, stratification, and maintenance defined by waste purification practices. Sigmund Freud detailed the anal stage as one of mastering bladder and bowel control – mastering in other words, the first waste of which we are conscious.[5] Wilk provides a contemporary version of this insight, arguing that human bodily waste "is contaminated through principles of *magic*, rather than chemistry ... just ask a North American undergraduate student to drink a glass of water taken from a brand new, unused, and thoroughly disinfected toilet. Or ask them to drink some of their own saliva" (2015, 226). In a sense, we continue this practice throughout our lives. Landfills swell with things we once wanted and now do not want; once valued and no longer value. What remains after our disgorgement is what we (want to) consider our real self. And, apparently, our real selves require not just consumption but extravagant overproduction and overconsumption.

Yet, as Rathje and Murphy argue, who we are is to be found in what we abject (2001; see also M. Douglas 2007). Skeptical of archaeology's ability to uncover in tombs, temples, and palaces anything other than what past peoples wanted to remember and be remembered for, Rathje and Murphy suggest that excavating rubbish sets the record straight. Rummaging through a person's waste – as any Hollywood stalker will attest – reveals intimate details of sexual practices, health (what we really eat versus what we report eating), personal hygiene, finances, personal relationships, private thoughts, political affiliations, and so on. Indeed, the Oxford English Dictionary defines garbology as the "study of human refuse in an attempt to divine ill-begotten truths" (in Spelman 2011, 315). As such, Spelman argues, "trash provides an epistemologically privileged resource for understanding what individuals and communities are all about" (2011, 323–4). In other words, we might come to better understand ourselves as Canadians if we take a critical look at the waste we produce… and attempt to disown.

### CANADA'S WASTE FLOWS IN BRIEF

Given the alarming statistics about how much waste, and the kinds of waste Canadians generate, it is entirely understandable that we would want to focus on ways we should be dealing with this waste. This is the crux of what is known in the business as "waste management" (WM). WM, as Canadians mainly encounter it, almost exclusively concerns dealing with waste once it has been produced. In other words, the waste we almost exclusively focus on is (1) the waste produced *post*-consumption, and (2) the waste produced by individuals and households. For instance, managing e-waste focuses on post-consumer e-waste – those iPhones, Samsungs, iPads, laptops, and desktop computers we no longer want – rather than the waste that goes into producing these and other objects. In fact, industries extract and consume an extraordinary array and volume of materials in fabricating these objects for our consumption. In his game-changing book *Reassembling Rubbish: Worlding Electronic Waste* (2018), Josh Lepawsky describes in detail how some 90 million pounds of e-waste is recycled through Dell's Reconnect program. This sounds like a lot until we learn that a single smelter operation in Mexico that produces copper and other metals used in electronics production produces 819,000 million tons of sulfuric acid waste. Lepawsky observes that the waste produced in extracting some of the materials used in electronics *from one smelter* is 1.8 times larger than e-waste exports *from the entire* United States of America.

Here we confront one of the little-known realities of waste in Canada: most waste is generated at the extraction and production stages rather than at the consumer stage. Let's take a moment to consider this, because this fact should profoundly shift our WM system. As reported in the 2012 edition of Statistics Canada's *Human Activity and the Environment* publication (which is our most recent statistical data source for countrywide waste production), the oil sands industry produced 645 million tons of waste (this is actually a 2008 statistic), the mining industry produced 217 million tons of mining waste and 256 tons of mine waste rock, the agricultural industry produced 181 million tons of livestock manure (this is a 2006 statistic), and households produced 34 million tons of MSW (a 2008 statistic). This means that industry (only including oil sands, mining, and agriculture) produced 97.4 per cent of our total waste. Households produced 2.6 per cent of our total waste. Now, if we figure in construction, demolition, and institutional waste (which includes places like universities, schools, and hospitals), then Canadian households produce significantly less than 2.6 per cent of Canada's total waste. Now let's factor in military waste. It is difficult (as section 2 of this book details) to obtain statistics on how much waste the Canadian military produces (on Canadian territory, let alone abroad). We have some information about historical waste sites, and the Canadian military publishes some data on how much waste is produced on military bases, but again this is institutional waste (i.e., food and so on) rather than the waste produced through military operations such as munitions testing. To include military waste in the equation, the 2.6 per cent figure needs to again be reduced considerably. So, although there is no way to obtain this data under the current data-collection systems in Canada, we may reasonably estimate that MSW accounts for less than 1 per cent of Canada's total waste production. Let us also further consider Max Liboiron's (2013) insight, that most municipal waste should actually be considered industrial waste that has been externalized to consumers in the form of packaging and disposable and poor-quality goods that individual households are then held responsible for sorting and recycling for free (those smart phones that mysteriously work less efficiently after only a few months), for waste and recycling companies to then buy back from municipalities, paid for by our taxes.

The latest *Environmental Indicators Report* published by Environment and Climate Change Canada (2018) does not disaggregate waste in the way that previous reports have. Under the banner "socio-economic indicators," the "solid waste disposal and diversion" category refers only to residential and non-residential MSW. These indicators, then, do not refer to industrial and/or military waste, nor municipal liquid waste. Nevertheless, the results point towards the same general findings: from 2002 to 2016 the total amount of waste collected in Canada increased by 11 per cent, or 3.5 million tons; and at least in 2016, the non-residential sector produced 59 per cent of disposed waste (Environment and Climate Change Canada 2018). This data is sourced from municipal governments, which only collect data on local businesses and households. Again, if we were to include industrial and military waste, this figure would be considerably higher. Other reports recently made public, such as Deloitte's bombshell report, *Economic Study of the Canadian Plastic Industry, Market and Waste: Summary Report to Environment and Climate Change Canada*, have found that 86 per cent of plastics end up landfilled, with only approximately 9 per cent of plastics being recycled (recycling, as chapter 2 will detail, is not a sustainable response to waste) (Deloitte 2019, ii). The Deloitte report also states that packaging, construction, and automotive waste accounts for 69 per cent of plastics waste.

These sobering statistics prompted the first of the two major arguments that I develop throughout this book: in Canada, the WM industry and the Canadian government (especially at the municipal level) purposefully configure waste discourse in ways that almost exclusively focus the public's attention on post-consumption waste (that is, municipal solid, and to a lesser extent liquid, waste) and individuals' responsibility for this waste. As chapters 2 and 3 explore in detail, most Canadians assume that our waste problems stem from individual consumption and not enough recycling, and the reason most Canadians assume this is because our participation in discussions of waste is governed in specific ways such that it begins and ends with individual responsibility and better technological innovations (better landfills, better EfW facilities, and so on). In this way, as Liboiron points out, "the individual rather than government or industry is represented as the primary unit of social change" (2010, 1).

Waste discourses are governed in this particular way because our nation is structured by neoliberal capitalism, which emphasizes a market economy, enhanced privatization, an overall decrease in government control of the economy, and a general entrepreneurial approach to profit maximization (Crooks 1993; Foote and Mazzolini 2012; for general discussions of governmentality see Burchell, Gordon, and Miller 1991; Foucault 1984b, 1988). The emphasis on individual responsibility operates within a capitalist rationale to *manage* waste in ways that do not disturb ever-increasing circuits of mass production and consumption (and therefore industry profit), producing an almost exclusive orientation towards downstream (disposal and recycling) responses to waste (Hawkins 2006; Kollikkathara, Feng, and Stern 2008; Lynas 2012).

Indeed, this shift from waste being a site of public contestation and protest in the mid-twentieth century to the largely uncontested contemporary management of waste speaks to the extraordinary success of neoliberal governmentality (Foote and Mazzolini 2012; Luton 1996; and more generally Beck 1995). Current Canadian WM legislation, policies, and practices speak to capitalism's production of excess – and "coping with excess is what passes in late-modern society for individual freedom" (Bauman 2001, 90). Further, WM practices such as recycling further shift attention from "industry and production to households and consumption" (Hawkins 2006, 104; Kollikkathara, Feng, and Stern 2009). A number of scholars have documented historical shifts in the governance of waste in North America from individual cartage and disposal operators to much larger corporations such as BFI and Waste Management Incorporated (Crooks 1993; Davies 2008; Melosi 2005). These publicly traded companies have invested heavily in all aspects of WM, including waste containers, collection and haulage vehicles, landfills, and other WM facilities, garnering enormous profits from municipalities that spend millions in contracts to these industries. In 2002, municipal governments spent more than $1.5 billion on WM services, and in 2014, Canada's WM industries earned nearly $7.1 billion (Statistics Canada 2017).

Thus, a political, economic, and cultural system dependent upon constantly increasing production and consumption, and on a stewardship approach to environmental issues, ensures that waste continues to be framed largely in terms of technological innovation, jurisdictions, and diversion practices produced through education, surveillance, sanction, and censure. As such, WM either capitalizes on or is victim to (depending on one's politics) what the Frankfurt School called "technological rationality," "in which the instrumental logic of rationalisation through 'technology' colonises every last aspect of modern life, including and especially thinking itself" (Grajeda 2005, 316). We see this almost every day, as politicians and industry representatives emphasize the need for technological innovations to solve our environmental crises – from global warming and fossil fuel dependence to species extinction and plastics recycling.

The upstream–downstream is a familiar metaphor within environmental studies and other cognate disciplines. This metaphor asks us to imagine a large industrial factory that is situated beside a river. The factory regularly dumps pollutants into the river, which travel downstream with the water current. A community of people lives some ways downstream from the factory. People in this community have higher levels of respiratory problems, cancer, neurological disorders, and the like. Doctors and other medical personnel do their best to treat these ailments, and the more affluent members of the community move to communities that are not situated downstream. For its part, the factory denies that its waste is causing human health issues and environmental degradation, and that scientists are working hard to determine causation, but this will require years of study and a lot of funding to complete the studies. Local and provincial levels of government maintain that they cannot intervene without scientific causal proof that the factory's effluent is causing the community's health issues. It is absolutely vital that communities living the most directly with waste contamination issues receive health care, but so long as the factory persists in disgorging its waste upstream, communities will continue to suffer downstream. Of course, this is less a metaphor and more the reality for many communities that are affected by the industrial production of environmental pollution (see, for example, Adeola 2012; Langston 2010; and Nixon 2011).

And so, in Canada as elsewhere, waste is largely understood as a technoscientific problem that requires technoscientific solutions (Gregson and Crang 2010; Lynas 2012). This points to a circular logic: engineering and science articulate the terms and parameters of waste problems such that each new problem tethers us to further solutions in the form of further technoscientific innovations. As such, most attention is directed towards more and better diversion (i.e., recycling), better landfilling and repository technology, and the development of new WM technologies. Yet our contemporary waste crisis involves the politics and economics of consumption, intergovernmental and industrygovernment relations, urbanrural divides, health, labour relations, gender and waste economies, sciencepublic relations, risk, governance, and so on (Maclaren and Nguyen 2003; Wynne 1987) – a bewildering array of factors, considerably beyond the remit of engineering and science.

Social sciences and humanities research – including environmental, communications, sociological, cultural, science and technology, philosophy, global development, geography, and political and economic studies – demonstrates the urgent need to better comprehend society's fundamental and inextricable entanglement with technoscientific phenomena, and in particular the indeterminate and contingent nature of their risks (Ali 1999; Bocking 2007, 2009; Coninck et al. 1999; Dodds and Hopwood 2006; Healy 2010; Petts 2001; Wynne 2006, 2007). This entanglement between technoscience and risk is also the site of Canada's present and future ethical responsibility to current and future generations. What former US Secretary of Defense Donald Rumsfeld (2002) called "unknown unknowns" – what we might call the uncertainties of the futurity of waste – draw our attention to the imprescriptibility of our responsibility to future human and environmental sustainability (Arendt 1958). If an engineering and scientific understanding of uncertainty is explained as risk that cannot be accounted for by probability, a socioethical understanding is formulated in terms that precisely acknowledge uncertainty itself as the basis of a social obligation toward the future. A landfill's contaminating lifespan is estimated at hundreds to thousands of years, and nuclear radiotoxicity endures for upwards of 100,000 years, or 3,000 generations (Beck 1992; Benford 1999; Benford et al. 1991; Rowe et al. 2004; van Wyck 2005).

In other words, while certainly necessary in the contemporary context, technoscientific ameliorations are downstream responses to emphatically upstream issues – issues, I contend, that require a socioethical framework that engages democratic deliberation, and action. This book seeks to examine the nature of the upstream (and therefore downstream) issues that come to the fore when the socioethical meanings of uncertainty are on equal footing with those of technoscience. As Harris Ali describes, this approach represents "the unsettling public realization that some modern environmental risks cannot be satisfactorily assimilated within the existent institutions of industrial society" (2002, 131). That is to say, there are risks that are beyond the control or definition of contemporary government and techno-scientific configurations.

Which brings me to my second major argument. Waste must be understood within the context of overarching upstream critical issues involving historical and ongoing settler colonialism, poverty, and racialized and gendered relations. When discussions of waste are isolated from profound forms of inequality, then it is far easier to maintain our focus on individual choices and individual responsibility. I argue throughout this book that waste is a profound and enduring *symptom* of inequality, and in Canada in particular, waste is a symptom of ongoing settler colonialism.

A number of studies focus on the association between waste and poverty, for instance where open dumps, landfills, and other waste repositories are proposed and sited (for example, Amegah and Jaakkola 2016; Furedy 1993; Mothiba, Moja, and Loans 2017; Parizeau 2006). Other studies (for example Dias 2016; Dias and Fernandez 2013) explore the ways in which people (and especially women and children) live on, and survive by picking through, dumps, in what Mike Davis describes as our "planet of slums," in his book of the same title (2007). In southern Canada, open dumps have been largely replaced by what are called sanitary (i.e., engineered) landfills. This is not the case, however, in Canada's Arctic, where open dumping is the norm. Many of the Arctic's original waste dumps were sited and developed alongside resource exploration sites in the early to mid-twentieth century, whereby sites were selected primarily for ease of access by short-term non-residents, including, significantly, the Canadian and American militaries (K. Johnson 2005); a resulting patchwork of contaminated and ill-designed dumpsites burdens already traumatized communities. These abandoned waste sites are a source of dangerous environmental contaminants. Among them are the Cold War–era radar stations that make up the Distant Early

Warning (DEW) Line, around which the soil commonly exceeds the 50 ppm of polychlorinated biphenyls (PCBs) allowable under the Canadian Environmental Protection Act (Stow, Sova, and Reimer 2005). PCBs are known endocrine disruptors and potential carcinogens. Because PCBs bioaccumulate and biomagnify in the tissues of wildlife (Braune et al. 1999; Giesy and Kannan 1998) and can leach distances of up to 25 kilometres (Pier et al. 2003), many Inuit country foods have become sources of contaminant exposure (Van Oostdam et al. 2005).

By and large, territorial governments want to adopt, and are expected to adopt, southern Canadian WM governance structures and technologies. Many Arctic communities, however, face unique challenges compared to their southern counterparts: community isolation increases the operational costs of WM facilities and can prevent the transportation of waste, recycling, and hazardous materials (Arktis Solutions 2010); almost all goods are shipped long distances, which results in increased packaging; a lack of trained personnel can hinder the proper operation and maintenance of WM facilities (Environment and Climate Change Canada 2017); and frequent extreme weather events can stall WM operations for significant periods of time. Beyond these practical material concerns are a plethora of economic and social conditions that can only make sense within the context of ongoing settler colonialism: food insecurity, housing shortages, suicide, drug and alcohol abuse, and a sharp lack of health care and other services.

## STUDYING CANADA'S WASTE ISSUES

For me, the topic of waste has become much like Alice's looking glass. The more I visit garbage dumps, EfW facilities, landfills, recycling hubs, and manufacturing industries in Canada and abroad; the more I observe people painstakingly trying to decide which bin to place their used coffee cup in; the more I talk to engineers, waste operators, government officials, food bank workers, environmentalists, industry representatives, and local residents living with contaminated environments; the more I watch various species of birds dive-bombing garbage; the more I immerse myself (mainly figuratively, but sometimes literally) in waste, the more keen I am to better understand how we ended up in such a literal mess, and how we might get out of it. This will be a never-ending journey for me.

Through serendipity or just plain luck, I have encountered immensely bright and inspiring people who share my interest in waste issues. I sat next to Peter van Wyck, a communications scholar at Concordia University who researches nuclear waste issues, on a flight between Saskatoon and Edmonton. This chance encounter led to our eventual work partnership on a number of grants that focused on waste issues. I am grateful to the Social Sciences and Humanities Research Council of Canada (SSHRC) for two Insight Development Grants and an Insight Grant, with which I have funded the case studies that illuminate the waste issues described in this book. I also secured ethics approval from Queen's University for all of the case studies recounted in this book. Around the time that I met Peter, I had just completed a year's work in an earth systems laboratory that culminated in a book on bacteria (*The Origins of Sociable Life*, 2009) and was deciding how I might follow a trajectory of working in an interdisciplinary fashion. I gave a presentation to a Queen's University senate committee about interdisciplinarity, and R. Kerry Rowe, civil engineer and landfill expert, chaired the committee. It was Kerry who first suggested that my interdisciplinary way of researching, and my interest in the inhuman, might be harnessed to study waste issues. Before my conversations with Kerry, I had not thought in any serious way about waste (beyond bacterial metabolism), so I certainly have Kerry to thank for introducing this topic to me, as well as for letting me visit his civil engineering laboratories and shadow some of his graduate students through their research projects.

I have always gained both the broadest and deepest understanding of a topic by using a variety of methods. I am equally comfortable (and uncomfortable) picking through an open garbage dump, interviewing a government official, reading newspaper articles about a trash crisis, and watching bottom ash being removed from an EfW facility. I have often been asked how I "gain access" to my field sites, and I tend to reply that, well, I ask. Truthfully, it is more than this. It's about *being interested*: spending time in laboratories, hanging out with scientists and engineers over coffee and lunch, attending seminars, looking over shoulders, opening refrigerators (asking first), talking to technicians, poking around – being there. Being there is crucial, and non-transferrable. It is about "having been somewhere," as Peter van Wyck learned in the *Highway of the Atom* (2010). One has to live with uncertainty to stand around, listen, observe, and ask questions. What sustains my willingness to live with this uncertainty in the research process is my abiding interest in what is going on. I am at times bewildered, frustrated, awestruck, ambivalent, and incredulous – but never bored. Here, for instance, is an excerpt from the field notes I took after asking Kerry if I could visit one of his landfill engineering test sites:

> The civil engineer meets me at the edge of the field. Here I am, in my steel-capped boots, fluorescent jacket, and hard hat. I have completed the hazards safety certification, and now I'm trying to remember the details of the hazards I need to watch out for. We walk up to the top of the hill, and the engineer draws my attention to various liners (different materials with different qualities); the angle of the hill in relation to the sun's rays; where the aerial blimp hovers overhead to take photographs; the network of liner wrinkles that stretch before us; the edges of the site where frogs and other small animals sometimes hop or slither onto the thick black liners and die in the heat of the sun; the workmen (they are all men) who are using some kind of machine specially designed to melt the liner edges together; men using a bigger machine to roll out liner like a carpet; the edges of the construction site towards the horizon, the owners of which have lent the engineer and his team this land on which to experiment; the trickle of water moving this way and that, finding its way to lower levels, and so on. I ask about nails, soil composition, the number of graduate students working here, the history of the site – anything I can think of to show that, while I may be naïve, I'm paying attention.

And what enables me to engage with scientific and engineering practices – and what I am really interested in, which is the matter they engage with – is a yearning for what Donna Haraway calls "response" (2008, 226), and the crux of this matter is that while we humans are representing and intervening, so too are all other critters, as, indeed, is the geologic (Hird 2012b). But I will readily say that I have never been more uncomfortable, less sure of my grounding, than when I have visited Canada's Arctic. It is in the communities of Iqaluit and Pangnirtung that I have most confronted, and been confronted with, my extreme privilege and ignorance as a white Canadian (see Simpson 2014; and Todd 2016). And it is this perpetual confrontation, and the discomfort that follows, that prompted me to realize, early on, that a book about Canadian waste issues *must* do more than just acknowledge the strong association between waste and colonialism.

The first section of this book, Waste in Southern Canada, focuses on waste issues in the most populated and privileged regions of Canada – that is, along our southern border with the United States. Chapters 2 and 3 emerged from the first SSHRC Insight Development Grant that I completed with Kerry. At that time, I was very fortunate to have two PhD students, Scott Lougheed and Cassandra Kuyvenhoven, who worked with me to understand how waste in southern regions of Canada is organized and governed. Soon afterwards, Alexander Zahara and Jessica Metuzals joined our team, followed by Diana Van Vlymen, Micky Renders, Hillary Predko, and Jacob Riha.

Chapter 2 is adapted from an article I published in 2014 with Scott Lougheed, Kerry Rowe, and Cassandra Kuyvenhoven entitled "Making Waste Management Public (or Falling Back to Sleep)." Scott Lougheed completed his PhD researching food recalls and waste (see Lougheed 2017). We were concerned with how and why publics assemble around WM issues. In particular, we explore Noortje Marres and Bruno Latour's theory that publics (different groups of the general public) do not exist prior to issues but rather assemble around objects, and through this assembling, objects become matters of concern that sometimes become political. The chapter addresses this theory of making things public through a study of Kingston, Ontario, where the landfill is closed, and waste diversion options are saturated. The city faces unsustainable costs in shipping its waste to the United States, China, and other regions. City officials are undertaking a cost–benefit assessment to determine the efficacy of finding a new site, or "siting," a new landfill or other WM facility. It was fortunate that Kingston was also the most easily accessible city for all of us, as we lived and worked there. This gave us the opportunity to "follow the story" of waste over a number of years (from 2012 onwards), attend town hall meetings, interview key informants, and steadily peruse the local media. This chapter emphasizes the complexity of making (or not making) landfills public, by exploring an object in action, where members of the public may or may not assemble, where waste may or may not be made into an issue, and where waste is so sufficiently made routine that it is not typically transformed from an object into an issue. Through this discussion, the chapter makes a strong case that waste in Canada needs to be the subject of sustained democratic participation and dialogue, rather than considered a technological issue that industry and government are best equipped to develop and resolve.

Chapter 3 is an adaptation of research that Cassandra Kuyvenhoven completed on a topic that Kerry Rowe suggested we look into – the issue of waste transportation. As Cassandra and I began to delve into this topic, we realized that the issue equally applies to recycling transportation. We again took Kingston, Ontario, as our case study because we had already collected archival and interview data on transportation (the topic had come up, even though we had not been especially concentrated on this topic at the beginning), as well as for the sake of feasibility, as both Cassandra and I were living in Kingston at the time. Chapter 3 critically analyzes what I argue resembles a "cult" of recycling in Canada. Canadian municipalities in southern regions of Canada, like those in other countries, all emphasize the 3Rs of waste: reduce, reuse, and recycle. Of these three, recycling is the least environmentally friendly but does not disturb (on the contrary, it encourages) circuits of capitalist production and consumption. This chapter argues that municipal governments in cooperation with industry successfully foster an "environmental citizenship" identity based on individual and household waste diversion (e.g., recycling), even though this accounts for a small percentage of the waste Canada produces. In so doing, members of the public are encouraged to survey and judge their own recycling behaviours as well as those of their neighbours, families, and friends, rather than the much more voluminous quantities (and often greater toxicity) of industrial and military waste. As such, this chapter argues that waste is an excellent example of neoliberal governmentality, in that the management of waste effectively controls populations, and diverts attention away from more salient upstream waste issues.

Section 2, Canada's Arctic Wastes, concentrates on the issues that waste raises in Canada's remote northern regions. Through my connection with Kerry Rowe, I was very fortunate to meet Allison Rutter, director of the Analytical Services Unit at Queen's University. Allison and her team spent some twenty years remediating various DEW Line sites across Canada's Arctic. After many hours of discussion about waste issues in the Arctic, I made my first trip to Iqaluit, Nunavut, with Allison in October 2013. On this trip, Allison introduced me to Nunavut, as well as to key members of the Nunavut Arctic College and Nunavut Research Institute, and we spoke with various government officials about waste issues. After this initial visit, I was very keen to return. What had heretofore been an interest amongst others became a keen interest to me: I began reading everything I could about Inuit *Qaujimajatuqangit* (Inuit knowledge) and Indigenous knowledge more generally (including that of Indigenous peoples of Canada, Australia, New Zealand, and Sweden), what non-Inuit often describe as "traditional knowledge." For instance, having formulated my thoughts around waste and settler colonialism, I found synergies with Glen Coulthard's *Red Skin, White Masks* (2014) around issues of the Canadian Crown defining "land" in treaty negotiations with the Dené. I found Belcourt's "Animal Bodies, Colonial Subjects: (Re)locating Animality in Decolonial Thought" (2014) important in thinking through the positioning of the inhuman (such as *tulugaq* – ravens – and *qimmiiq* – sled dogs) by Inuit and by Canadian authorities (the RCMP). I also found McCluskey's compendium of Inuit oral stories of *tulugaq*, *Tulugaq: An Oral History of Ravens* (2013), vital for better understanding human and inhuman relations of kin, labour, and intergenerational continuity.

Like Canadian southerners before me, I was shocked at the amount of visible waste in Arctic communities. I remember that as a child my father used to take my sisters and me to one of the open dumps in Ottawa. It was a sort of family weekend ritual to drive to the dump to watch the bears scavenging. Ottawa's open dumps have long since disappeared, but here they were (and then some) in Canada's Arctic, our "true north strong and free." So, on one of my next visits, I took my new master's student, Alex Zahara. We walked around all the known dumpsites (as chapter 4 details, there are several) and talked to as many researchers, government officials, business owners, and community residents as were willing to talk with us. It was on this particular trip that, on one of my many walks around town, I stumbled on a group of men concentrating on a pipe coming out of a sewage truck. I ended up having a lengthy conversation with these rather bewildered but nevertheless polite men (why was this privileged white woman from Kingston, Ontario, so interested in sewage?), and accompanying them to Iqaluit's open sewage lagoon, where I watched them disgorge the contents of the sewage into the open lagoon that overflows into Frobisher Bay.

Chapter 4 provides an examination of waste issues in Nunavut's capital and largest community. In the Eastern Canadian Arctic city of Iqaluit, a three-storey pile of waste rests atop a peninsula that extends well into Frobisher Bay. The dump is one of several waste sites in Nunavut's only official city. It joins an unknown number of waste sites that the US and Canadian military, oil, gas, and minerals industries have left abandoned on, and in, the landscape. This chapter examines waste within the wider context of settler colonialism, as well as contemporary neoliberal governance practices, to argue that waste is itself part of a colonial context within which Inuit and other Indigenous peoples in northern Canada continue to live. Waste is a provocative material concept with which to think about ongoing settler colonialism. Whether in the form of mining, nuclear, industrial, hazardous, sewage, or municipal, and whether it is dumped, landfilled, incinerated, or buried deep underground, waste constitutes what will likely be the most abundant and enduring trace of the human for epochs to come. This chapter takes up the challenge posed by Dipesh Chakrabarty (2009) to conceptualize the colonial subject, wherein humanity is re-characterized as a geophysical force. This conceptualization speaks of a globalized human race to whom past and present generations project responsibility and reparation. This chapter will emphasize how the effects of this waste landscape – neocolonialism's dividend – is experienced differently by Inuit people living in Canada's Arctic.

My conversations with Allison Rutter about her DEW Line remediation project, and our visit to one of the sites that her team remediated (the radar station in Iqaluit), led me to the scientific and engineering literature on contaminated sites cleanup. My research interest grew into a project in its own right. Chapter 5 focuses on Arctic military waste legacies. During the Cold War, the United States and Canada embarked on an ambitious military construction project in the Arctic to protect North America from a northern Soviet attack. Comprised of sixty-three stations stretching across Alaska, Canada's Arctic, Greenland, and Iceland, the DEW Line represents both the largest military exercise and the largest waste remediation project in Canadian Arctic history. Despite the massive cleanup operation undertaken, the DEW Line's waste legacy endures as a prominent and deeply rooted feature of Canada's Arctic history. This chapter explores waste as a key issue in the shifting narratives around the modernization of the Canadian Arctic. While the DEW Line has been extensively analyzed in terms of its effects on the modernization of the Arctic, this chapter seeks to link Canadian sovereignty, security, resource exploitation, environmental stewardship, and Inuit self-determination directly to waste issues. As industrial activity and military exercises stand to significantly increase in the Arctic, I draw attention to the lessons of the DEW Line – that a philosophy of "develop now; remediate later" incurs steep human health, environmental, financial, and political costs.

On my initial visit to Iqaluit with Allison, and my subsequent visits with Alex and then with my children, and then with Micky Renders, I noted with fascination the ubiquity of the ravens that not only scavenge at the open garbage dumps, but also perch on the buildings, fly through the streets, and generally make themselves known (and heard) in Iqaluit. On my initial drive out to the West 40 dumpsite with Allison, I became absorbed by the sled dogs tethered at the side of the dirt road past the military base and airport. Indeed, I walked back to these dogs several times on my various trips and, keeping a safe distance, watched ravens playing with (i.e., taunting) the chained sled dogs. All this time, I was immersing myself in Inuit writing and song, some of which detailed the importance of various birds and animals in Inuit cosmology and experience (see, for example, McCluskey 2013; d'Anglure and Levi-Strauss 2018; Laugrand and Oosten 2010; Briggs 1999; Tagaq 2014). This literature and song, the things I witnessed, and my longstanding interest in the inhuman culminated in an examination of the Arctic's scarcely considered waste legacy – outlined in chapter 6 – which I convinced Alex to work on with me while on our visit to Iqaluit. As an unintended (and often unacknowledged) fallout of capitalism, we have developed sophisticated technologies to hide our discards: waste is buried, burned, gasified, thrown into the ocean, and otherwise kept out of sight and out of mind. Some inhuman animals seek out and uncover our wastes. These "trash animals" choke on, eat, defecate on, are contaminated with, play games with, have sex on, and otherwise live out their lives on and in the waste of our formal and informal dumpsites. In southern Canada's sanitary landfills, WM typically adopts a "zero tolerance" approach to trash animals. These culturally sanctioned (and publicly funded) facilities practise diverse methods of "vermin control," such as the sanitary landfill I toured that services Montreal, where hawks are enlisted to keep gulls from feeding from the garbage. By contrast, within Inuit communities of the Eastern Canadian Arctic, ravens eat in, play in, and rest on open dumps by the thousands. Chapter 6 explores the ways in which Western and Inuit cosmologies differentially inform particular relationships with the inhuman, and trash animals in particular. Alex and I argue that waste and wasting exists within a complex set of historically embedded and contemporaneously contested neocolonial structures and processes. Canada's Arctic is a site where differing cosmologies variously collide, intertwine, operate in parallel, or speak past each other in ways that often marginalize Inuit and other Indigenous ways of knowing and being. Inheriting waste is more than just a relay of potentially indestructible waste materials from past to present to future: through waste, we bequeath a set of politically, historically, and materially constituted relations, structures, norms, and practices with which future generations must engage.

Sections 1 and 2 of this book provide theoretical and empirical evidence concerning the state of WM in southern and Arctic regions. In section 3, Waste at the End of the World, I move the discussion to a more theoretical register that engages with both the indeterminacy of waste, and waste as a physical signature of the Anthropocene. This shift creates an important shift in scale, from local actions and contexts to global effects. Chapters 7 and 8 demonstrate that waste is not only a *global* concern that cannot be meaningfully understood outside of social justice concerns, but it is a *planetary* concern. That is, whereas the chapters in sections 1 and 2 focus on the political, cultural, economic, and social forces that produce our current waste crisis, chapters 7 and 8 demonstrate that waste is also *material*, and this materiality conditions how waste itself engages with the planet. I explore this argument by showing how, despite efforts to control waste (through forms of governance, including settler colonialism), it nevertheless proves to be rather stubbornly indeterminate (as civil engineering and biological studies of leachate, toxic compounds, and so on detail). Chapter 9 then returns to the theme of governance to show (through Alex Zahara's empirical case study – see Zahara 2018) how local residents in Iqaluit were able to harness the indeterminacy of waste to convincingly argue against the Canadian federal government's governance of the West 40 dump fire. As such, chapter 9 brings full circle the focus in section 1 on waste governance in southern regions, the focus in section 2 on waste governance in the Arctic, and the focus in section 3 on the material limits of the political, cultural, economic, and social governance of waste.

Chapter 7 connects waste issues to discussions about the Earth's strata, and their human and inhuman perturbations, arguing that we want to be very cautious when generating waste from techno-scientific processes such as hydraulic fracturing and mining. This chapter is one outcome of over twenty years of discussion with Nigel Clark, whom I met as a colleague at Auckland University. I have lost track of the number of slow walks I have shared with Nigel as we've discussed my fascination with and deep respect for bacteria, and his reverence for inhuman planetary forces such as fire. When my research focus turned to waste some years ago, it was an easy segue to examining bacterial metabolism in the garbage we store underground. Microbes famously starred in Latour's (1993) pathbreaking account of a modern networked power that hinged upon turning microscopic life into visible, present, and negotiable participants in socio-political worlds. Microorganisms continue to feature in accounts of global power in which human actors mobilize – globally, speedily, even preemptively – to counter threats of emergent pathogenic life. But in such framings of space, power, and inhuman agency, there is still a propensity to analyze as though we are only dealing with the surface of the planet and then to complicate these – rather than setting out from worlds that are always already volumetric, stratified, and deeply temporal. The challenges – and opportunities – associated with bacterial life arise out of the fact that microorganisms are a condition for the possibility of more complex life, ourselves included. Chapter 7 explores a number of contemporary practices of waste disposal in landfills, which sooner or later serves to supply heterogeneous worlds of subsurface bacteria with nutrient-rich broths of leached liquid. Through these examples, the chapter develops the idea that bacteria bring into relief the way that human power relations operate with and through other life forms. More than this, the subtending relationship that bacteria have with us – that is, that all life on earth was generated by microorganisms, and that all life continues to be dependent on microorganisms – offers an opportunity to consider how power relations need to be thought of not only in terms of spatial networks, scalar leaps (which is to say, thinking conceptually at different scales), and the enfolding of topological layers over previous layers, but also by way of complex, multidimensional negotiations between different spatio-temporal strata.

Chapter 8 connects waste issues to current broad, theoretical interest in speculative realism, feminist materialism, and antiracist theories. From a feminist materialist perspective, careful attention must be paid to the work involved in material determinations – and specifically, to how that work delineates what becomes important, ontologically and epistemologically. By contrast, the speculative turn in philosophy draws our attention to the limits of a knowledge that is based upon an experiencing (human) subject, and to the importance of considering the autonomy of objects and the limits of relationality – or our relationship with these objects. This chapter links this varying emphasis on autonomy or relationality to waste contamination and environmental politics in the Anthropocene. The contamination lifespan of a landfill is estimated at hundreds to thousands of years, and nuclear radiotoxicity endures for upwards of 100,000 years, or 3,000 generations, making the consequences of the re-stratification of waste imprescriptible, and a difficult material and cultural basis upon which to determine a social obligation to the future. In contrast to the extinctions foregrounded in the Anthropocene (including large fauna such as polar bears and tigers), the accumulations of bacterial strata in dumps and landfills promise a generative life underground that interacts with – "speculates" with – the materiality of waste. Withdrawn from sight and sensibility, such generative progeny suggests the inherent *in*determinate nature of the material-discursive practices of human strata and the impossibility of predicting, let alone controlling, what bacteria will create with our waste. This chapter argues that any recourse to relationality – to seeing ourselves in relation to material waste – needs to attend to the autonomous and creative or speculative life of the materials that subtend it. I am particularly indebted to Nils Johansson's PhD research on "urks" – the discards of urban infrastructure – in writing this chapter.

Chapter 9 brings these theoretical discussions back home, with a focus on the question of the futurity of waste and our ethical responsibility to past, present, and future generations. Iqaluit became a centre of controversy when its main West 40 dump spontaneously caught fire on 20 May 2014. This chapter argues that Iqaluit's "dumpcano" (as it came to be known) may be usefully understood as a virtuality that temporarily condensed a set of relations between the dump as a material object (or more specifically a multispecies biogeology) and a number of economic, cultural, political, and social conditions. As it happened, with the support of my SSHRC grant, I had sent Alex Zahara to live in Iqaluit during the summer months of 2014. So, Alex was there during some of the dump fire controversy and was able to attend town hall meetings, talk to local residents, and interview key people like the chief health officer and the fire chief. Alex ended up taking his own equally interesting approach to the data analysis (see Zahara 2018). Chapter 9 details my take on what happened during this controversy, and examines how scientific and government discourses attempted to convey to the public a uniform message of *scientific certainty* concerning the levels of contamination and threats to human health and the environment. For their part, concerned residents and emergent activist groups engaged with official and unofficial messaging in terms of *material uncertainty*. As such, the discourses that developed around the dump fire made visible and helped register, or obscured and attempted to displace, in different ways, the shifting material properties of the dump. The chapter makes the case that (government-defined) non-expert residents were able to effectively mobilize scientific uncertainty to draw attention to the links between the dump fire and issues of Indigenous social justice.

And finally, the epilogue focuses on uncertainty as the basis for discussions about the futurity of waste and our ethical responsibility to future generations of Canadians and other nations, as well as to the environment and inhuman living organisms. Not only can we not fully grasp the hazards of past and present waste issues, but future waste issues will always necessarily remain even more provisional and speculative. What does it mean to acknowledge uncertainty as an ethical basis from which to consider waste – and to consider it as a phenomenon about which we are not simply obligated to warn future generations but with which we are to intervene through democratic deliberation and action? To wit, the epilogue argues that any form of intervention must resist the tendency to focus on downstream responses, and instead tackle definitively upstream issues such as national and global consumption within global neoliberal capitalism. Indeed, the epilogue considers what it would mean for Canada to become a global leader in upstream approaches to the futurity of waste.

CHAPTER TWO

# Southern Canada's Waste Management Problem

INTRODUCTION

Given the vast regulatory, engineering, transportation, science, policy, governance, behavioural, and other considerations necessary to maintain our Canadian waste management (WM) system, it is remarkable that waste is, for the most part, so unremarkable. Like the "silent workings" of the sewage system in Paris, WM in Canada has, for the most part, "become part of the daily routine of administration and management" (Latour 2007b, 817). Unless there is a change or proposed change, or unless there is some form of temporary crisis, such as Canadian waste ending up in the Philippines, WM is typically so routinized that it does not garner or sustain the public's attention as an issue, and thus does not become political. As Noortje Marres succinctly puts it, "no issue, no politics" (2005a, 62). As I will argue in this chapter, "no politics" happens when "objects" – something constructed as political in its widest sense – such as WM are governed in such a way that they do not engender public interest. Michel Foucault analyzed this process as "governmentality"; more recently, Latour refers to the process as "Politics-5" (Foucault 1991; Latour 2007b). Both theorizations forefront the role of power in defining and framing WM as an issue or a non-issue. And so, even when we do focus on waste, we almost exclusively concentrate on the smallest category of waste produced: that by Canadian individuals and households. It is important, then, to understand how waste either becomes political or is prevented from becoming a political issue.

As waste researchers convincingly demonstrate (for example Ali 1999; Dodds and Hopwood 2006; Healy 2010; Petts 1998, 2001), waste garners public interest and becomes an issue when existing landfills erode, explode, leak, or slide and thereby compromise human health and/or the environment. Waste is also brought into view by garbage collection strikes, new landfill sitings, non-divertible waste technology proposals (including waste-to-energy technology), and so on. As such, waste in Canada does not become an issue unless it is brought (back) into view and members of the public invest it with particular meanings. These meanings have to do with known, unknown, and unknowable risks; health; consumerism; trust in science; property values and taxes; labour; environmental justice; and so on. Waste in southern regions of Canada, in other words, requires work to become a political issue. This work, generally an ongoing activity involving human and inhuman relations (Spector and Kitsuse 1972), involves various mechanisms – exploding landfills, feasibility studies, town hall meetings, and so on – to construct WM as either an issue or a non-issue. Existing power relations help create a framework for these mechanisms, which will determine, to the fullest extent possible, whether WM will become a political issue or not. And even when waste does become an issue, it only tends to become a matter of concern for short periods of time: "To make a thing public is only a moment in the life of an issue, an intense and uncertain episode to be sure, but neither its first nor its most final" (Latour 2007b, 818). Moreover, *not* having to participate in any given matter of concern for which action is required, Latour argues, is the most prevalent public response. Understanding how waste becomes political, in other words, is as much about how publics "fall back to sleep" as it is about the public's mobilization to action (Latour 2007b, 819).

This chapter explores WM as a site of neoliberal capitalist governmentality, or as Latour would have it, Politics-5. This form of governance leads to the configuration of WM as a technological issue supported by norms and practices of individual responsibilization. That is, WM is largely structured as a matter of responding to individual citizens' waste "needs" through industry and technology, rather than, for instance, as a socio-ethical issue requiring forms of democratic deliberation, on such issues as our reliance on an economic system based on relentless growth, the dependence of government on industry, and overproduction and overconsumption. As chapter 1 outlined, framing WM as a technological issue circumscribes discussions to focus on better WM technologies (longer-serving landfill liners, better ways of disposing of incinerator fly ash, and so on) and diversion (primarily recycling), the latter for which members of the public are largely held responsible. This articulation of governing, then, makes use of a particular public in relation to waste, one that conceptualizes waste at the individual level, to be resolved with downstream techno-scientific innovations and more responsible individual behaviour. This in turn leads to differential assessments of WM risk among scientists, members of the public, community group members, government officials, and so on. From this starting point, I examine the processes through which waste does and does not become an issue, creating or not creating a concerned public that engages in time-limited, constrained responses. Using a number of Canadian examples, this chapter examines the processes through which an apparently mundane and unremarkable object becomes an issue and then a matter of concern, how it attracts attention, and gathers momentum and calls for action – or doesn't. With Marres (2007), I argue that issue formation and dissolution is a critical dimension of the democratic process, and one to which we need to pay more attention.

## MANAGING RISK

Risk assessment has become a prominent contemporary response to the techno-scientific complexities of the management of waste. Risk assessment emphasizes known risks at the expense of unknown risks, and shifts attention from prevention to permissible levels of contamination (Hale and Dilling 2011). And thus, when municipal governments undertake new landfill siting assessments, or are dealing with soil and water contamination by landfill leachate – and its negative health effects on humans, animals, plants, the soil and so on in the landfill vicinity – members of the public encounter what it means to live in a risk society. The extensive risk literature (see Beck 1992) notes that members of the public are increasingly being asked to deliberate about environmental, health, and other issues for which the risks are inherently indeterminate (MacFarlane 2003). Issues such as contaminants of emerging concern are somewhere between what Donald Rumsfeld called "known knowns," "known unknowns," and "unknown unknowns." Contemporary risk society is characterized by diversified publics that know that inherent risks accompany what politicians and industry term "technological innovation" (waste-to-energy technologies and so on), and are increasingly wary of immediate and short-term risk assessments that underplay the indeterminacy of the risks that accompany technologies – the "unknown unknowns." Indeed, as Michel Callon (1999) argues, non-specialist members of the public make a rational decision *not* to trust governments who do not address the indeterminacy of risks that are endangering society.

Syed Harris Ali, drawing on Ulrich Beck's well-known risk society theory to understand how residents of the city of Guelph in Ontario, Canada, responded to a landfill siting exercise in their community, notes that, with regard to landfills engineered prior to the 1990s, "the concentration of substances in landfill-generated leachate that can cause death may be parts per billion, that is, at a concentration that cannot be tasted or smelled in the drinking water" (1999, 4). Ali draws attention not only to the risks known to engineers and scientists, of which members of the public have no direct perception, but also to risks the public may see as unknown and unknowable to engineers and scientists. Compounding the problem, these risks are often explained to the public in terms of *a priori* acceptable levels of negative impact. As Beck notes, "The really obvious demand for non-poisoning is rejected as utopian. At the same time, a bit of poisoning being set down becomes normality. It disappears behind the acceptable values. Acceptable values make possible a permanent ration of collective standardized poisoning" (Beck 1992, 65; see also Langston 2010).

Members of the public, such as residents of Iqaluit, as chapter 9 will detail, are wary of scientifically described risks – which are, by definition, known risks typically involving numerical thresholds of acceptable environmental compromise – and want to focus on why some health and environmental risks are deemed acceptable, as well as on unpredicted effects, which scientists cannot address (Wynne 2006, 216).

## MAKING THINGS PUBLIC MEANS BRINGING THINGS AND PEOPLE TOGETHER

Writing in the 1920s, Walter Lippmann recognized that the world had become too multifaceted for people to grasp all of its complexities. Nevertheless, he wrote, it is precisely within the context of complex problems that public determination of solutions is required:

it is controversies of this kind, the hardest controversies to disentangle, that the public is called in to judge. Where the facts are most obscure, where precedents are lacking, where novelty and confusion pervade everything, the public in all its unfitness is compelled to make its most important decisions. The hardest problems are those which institutions cannot handle. They are the public's problems. (1993 [1927], 121)

It is within the context of risk management that Canadians are increasingly called upon to consider and ultimately accede (because waste must go somewhere) to the adoption of a landfill and/or other WM technology through public consultation exercises. (Alarmingly, exported waste never receives this type of scrutiny unless it is radioactive.) So, let's take a look at how WM issues are brought to the public's attention.

Canadians are typically exposed to WM in terms of the mundane practices of sorting through household waste, depositing waste and diversion material for curbside pickup, and occasionally divesting waste through transportation to specified hazardous waste depots, yard sales (in which case, waste transforms into a resource), and the like. In other words, this form of governmentality focuses on individual attitudinal and behavioural responsibilization rather than deliberate larger upstream issues, such as the association between economic growth and waste (see Schnaiberg 1980).

Waste tends to shift from being a mundane object to an issue when municipalities consider increased user fees for waste disposal or declare the need to site a new landfill and/or introduce another WM technology such as an incinerator. This shift may also occur when landfill leachate breaks free of its constraints, particulates and organic compounds from incinerators infiltrate human lungs, composting sites emit nauseating odours, bioreactors malfunction, or the masses of waste necessary to "feed the beast" of incinerators leads to the importing of other municipalities' (or countries') waste.[1]

As such, WM processes, and our relationship with waste more broadly, must typically be "made manifest or actually materialized in some way, so that they become visible and others can be convinced of their potency" (T. Clark 2012, 20; see also Hannigan 2006). And as section 3 of this book will detail further, emerging publics are assembled not only around arguments, values, and interests, but out of combinations of heterogeneous materials, processes, and inhuman things such as dump smoke, dogs, and birds (Mahoney, Newman and Barnett 2010). As Latour notes, "every new non-human entity brought into connection with humans modifies the collective and forces everyone to redefine all the various cosmograms" (2007b, 816). Indeed, when waste becomes a matter of concern it brings disparate rights holders and stakeholders together: politicians and waste ash, engineers and conservation, non-governmental organizations and Facebook, radio broadcasters and toxic chemicals, Indigenous communities and social scientists.

### POLITICS AND POWER

In his article "Up and Down with Ecology," Anthony Downs (1972) describes an "issue-attention cycle" and the stages through which some objects become issues (his "pre-problem" and "alarmed discovery and euphoric enthusiasm" stages), others may become framed in ways that set the parameters for solutions (the "realizing the cost of significant progress" stage), and others may serve to decrease public interest as solutions appear unrealizable, appear to threaten lifestyles, or appear undesirable (Downs's "gradual decline of intense public interest" and "post-problem" stages). How an object moves through these various stages is, as Downs notes, a function of power (see also Hannigan 1995). This is particularly evident once an object has become an issue of concern and people begin to realize that solving the problem would mean, for example, a significant reduction of, and reorientation to, consumer lifestyles, and challenging government and industry, which requires significant energy and time, and often money. This would be the case, for instance, if Canadians generally became aware of the relatively insignificant amount of waste that households produce compared with that of industry, as well as the significant limits of recycling as a WM solution (as detailed in chapter 3).

More recently, Latour (2007b) identifies five ways in which politics may assemble around objects. First, things may become political when a new entity (such as leachate, or fly ash) is brought into connection with humans such that it modifies and forces them to redefine the collective – that is, when the entity, as John Dewey (1954 [1927]) observes, exceeds the procedures of institutional politics. A second, related form of politics emerges when an issue "entangle[s] many unanticipated actors without [experts] having developed ... instruments to represent, follow, take care of, or anticipate those unexpected entanglements" (Latour 2007b, 816), generating an unsettled public. A third form emerges when a government attempts to frame an issue in terms of a clear general will or common good, and fails. As analyses of various WM siting assessment exercises throughout the world attest, WM industries operating in tandem with municipal governments increasingly ask members of the public to accede to formulaic assessment exercises that circumscribe the parameters, for example to discussions of disposal technologies and increased recycling. Once this major parameter is set in advance, discussions are further circumscribed to a limited number of consultation and discussion events, and their time frames (Ali 1999; Coninck et al. 1999; Dodds and Hopwood 2006; Einsiedel, Jelsøe, and Breck 2001; Healy 2010; Petts 1998, 2001). Power circulates through this framing, which is increasingly managed by multinational corporations specializing in waste technology assessment, siting, construction, operations, monitoring, closure, and aftercare. With on-site engineers and scientists, networking with government, and sophisticated, well-budgeted, in-house public relations management teams, these new brokers increasingly manage municipal and public discussions of WM through feasibility reports, town hall meetings, presentations, and other forms of consultation (Allen 2007; Corse 2012; Marres 2005b, 2; Van de Poel 2008). Indeed, neoliberal capitalist governance enhances industry's monopoly by embedding techniques such as public consultations and feasibility studies within industry's remit. Latour's third sense of politics emerges when these techniques fail – when, for instance, members of the public become skeptical that they are not getting the full story.

A fourth politics emerges in what de Vries (2007) describes as the deliberation of "mini-kings," or when "fully conscious citizens, endowed with the ability to speak, to calculate, to compromise and to discuss together, meet in order to 'solve problems' that have been raised by science and technology" (Latour 2007b, 817). An example is the citizens of Ryedale, in North Yorkshire, UK, who met with engineers and scientists to resolve the issue of recurrent flooding in their community (see Whatmore 2009; Lane et al. 2011; Landström et al. 2011).

A fifth form of politics, Politics-5, emerges from objects that are so naturalized that they do not appear to raise issues: "all those institutions [that] appear on the surface to be absolutely apolitical, and yet in their silent, ordinary, fully routinized ways they are perversely the most important aspects of what we mean by living together – even though no one raises hell about them and they hardly stir congress[people] out of their parliamentary somnolence" (Latour 2007b, 817). Latour identifies Foucault's theory of governmentality with this form of politics. Current neoliberal governance makes WM appear routine and apolitical by adopting a discourse of "efficiency" and "cutting red tape," refusing to fund expert third-party review or hold public hearings where all the issues can be raised and deliberated, cutting the available technical expertise amongst its own regulators, and reducing approvals to a narrow bureaucratic process rather than one that seriously considers broad technical and social issues (see chapters 3 and 9 for Canadian case studies that detail the use of exactly these techniques). Politics, then, may equally turn on issues returning to objects as people "fall back to sleep," as on "convening, mobilizing, and sustaining a public" (Latour 2007b, 817).

In Canada, it is a legal requirement that new landfill siting and the adoption of other WM technologies be made, to some degree at least, public. And this process is important. As Canadians, we want to better understand the operations of power through which municipalities identify, present, and thereby attempt to stabilize an object in particular ways (DeSilvey 2006); which authorities (industry representatives, engineers, scientists, government officials, and so on) are gathered in assessment exercises, and how these exercises are defined (Latour 2005); when and how members of the public are invited or not invited to participate in various kinds of public assessment consultations (Coninck et al. 1999; Einsiedel, Jelsøe, and Breck 2001; Goven 2003, 2006); and how these processes attempt to create a certain kind of member of the public (such as the "consumer" and the "good citizen").

## POLITICS IN THE MAKING?

WM constitutes an emerging issue in the small city of Kingston, Ontario, where I work. Scott Lougheed, Cassandra Kuyvenhoven, and I spent several years studying Kingston's past and present WM practices and the claims being made about these practices. We interviewed twenty-six key informants (this number is increasing as I continue to discuss WM issues with various rights holders and stakeholders), including six elected members of Kingston's city council whose remit includes WM and whose constituencies cover the areas of Kingston (where previous landfills have existed or where new landfills have been sited and failed to gain approval); the managers of Kingston's main waste processing centre (the Kingston Area Recycling Centre, or KARC, as it is locally known), who are accountable to the city council and responsible for contracting out Kingston's waste to various private companies; representatives of various community groups concerned with environmental issues, including WM; and two WM industry representatives. We also gathered observational data from waste processing centre tours and town hall meetings, and used archival sources of data, including WM industry documents commissioned by Kingston City Council, documents from the WM processing centre, and city council web-based documents available to the public, and approximately eighty newspaper reports concerned with past WM issues in Kingston. We also constructed and administered a web-based survey to 107 Kingstonians to determine the willingness to participate in political deliberations on the future of WM in the city, such as on waste minimization policies (for example, limits on the number of bags a household can put out for curbside pickup), as well as on potential new waste processing facilities to replace a recently exhausted landfill. Our study identified four key stakeholder groups: municipal government representatives (city councillors, City of Kingston staff members, and others); industry; media; and individual citizens, or "the public." We analyzed how each of these stakeholders manages waste as an object of potential concern, beginning with the municipal government representatives.

To situate this case study, in 2017 Kingstonians generated approximately 41,795 tons of municipal solid waste (MSW), 16,405 tons of which were landfilled and 25,355 of which were diverted mainly through recycling programs (Utilities Kingston 2019). By 2031, conservative estimates are that the annual tonnage of waste will increase to 61,636. In 2012, the city paid Waste Management of Canada, a private company, $1.9 million to "handle and dispose of" residual MSW (City of Kingston 2012a, 23). This was up from $1.39 million in 2008 (City of Kingston 2009). The city spends approximately $6 million on WM, factoring in additional services such as $1.7 million in recycling services and $1 million on green bin services (organic waste) (Schliesmann 2012a). In 2002, Kingston City Council sought to improve its 38 per cent diversion rate – which refers to how much waste is taken out of the waste stream through reduction, reuse, and recycling – and its WM services in general in order to "protect the health, safety and natural environment of our citizens through fiscally responsible ... practices that encourage waste reduction and recycling, and that promote economic prosperity" (City of Kingston 2002, 4). The city council established several criteria for gauging the effectiveness of its strategy: increased diversion rate, reduced waste generation, and positive perceptions among citizens of city spending on WM services (City of Kingston 2002). Consistent with Latour's Politics-5, in which government and industry forefront a discourse of "efficiency," these effectiveness measures were further qualified by certain "efficiency measures," namely, low operating costs for the collection and disposal of waste and diverted materials. Despite the city's stated concern for the "natural environment," none of these indicators involved direct observation of environmental impact.

The primary means by which the city anticipated improving its "effectiveness measures" was to "rely heavily on our ability to educate our residents and encourage them to change their usage patterns and behaviours" (City of Kingston 2002, 4–6). From the outset, then, the city's approach to WM operated within a governmentality framework whereby responsibility for waste is assumed by individual citizens, brought about through adopting a deficit model, which theorizes that the general public lacks information and, if given sufficient correct information, will change their behaviour. Over several years, the city introduced grey (paper), blue (glass and plastic), and green (composting) bin diversion, and most recently adopted a one free garbage bag policy (MacAlpine 2012). Nevertheless, in January 2011, the city's landfill, now full, closed to the public. As a result, 100 per cent of the city's non-diverted waste is currently exported to other regions, including the United States. Private WM firms that transport the city's waste are retained on five-year contracts (Kingston City Council 2013a).

In a move that follows exactly the type of formulaic process this chapter examines – Latour's Politics-5 – Kingston city councillors in 2006 commissioned WM consulting firm Jacques Whitford (later replaced by another consulting firm, HDR) to produce an integrated waste management (IWM) study, at which point the city announced its goal of becoming "Canada's most sustainable city" (City of Kingston 2012b). The IWM study would consist of a twenty-five-year WM plan and recommend a suitable waste processing facility for the city that would reduce the city's reliance on transporting waste long distances and on landfilling (Jacques Whitford 2008a, 2008b).

Concerns about the cost of deploying new WM technologies and the uncertainty of maintaining Kingston's current status quo (of waste export) serve to link non-diverted waste, as an object, to issues (increasing taxes and fees, and environmental concerns about transporting waste to other regions), and potentially to politics. Many city councillors favoured a local waste processing facility, as they deemed transporting waste to out-of-town landfills to be contrary to their sustainability goals (Kingston City Council 2012). Some councillors argued that just because finding a waste processing technology was the "next logical step [it] does not infer that it is a pressing priority ... and that maintaining the status quo is an option" (Kingston City Council 2012, 7). This suggests, in line with Latour's fifth sense of politics, that the transport and landfilling of waste had become so routine – entrenched in city budgets and infrastructure, and with no immediately apparent environmental consequences for the city (consequences that the city staffers had not examined) – that it had yet to be seen as an issue for certain members of city council. For detractors on city council, however, the proposal of a new waste processing facility shifted WM from an object to a recognized issue. In addition to concerns about new and untested technologies or technologies that require a steady stream of waste to operate efficiently (like EfW facilities), councillors were concerned about investing heavily in a business that might eventually fail (Kingston City Council 2012). Despite these concerns, proponents argued that a local waste processing facility could allow the city to "turn garbage into a resource" (ibid., 8), which as one councillor argued, is a prerequisite for all "sustainable" cities. Turning waste into a resource is seen as more environmentally – and financially – sustainable in the long term, despite the initial outlay of capital when compared to maintaining the status quo of transporting and landfilling, in which there is no possibility of recovering WM costs.

By the end of our study, city councillors had not reached a consensus about precisely how the issue should be dealt with. In a manifestation of Latour's Politics-5, there was a general concern within city council to prevent politics from assembling around the issue of WM. The appearance of WM as apolitical is maintained in large part by engaging the public as participants in downstream responses through avenues that circumscribe the terms of public engagement. As the next chapter will detail, this is achieved by creating particular identities of what a "good citizen" is through inducements to recycle and compost (City of Kingston 2013f) and by minimizing the overall impact on individual lifestyles and consumption rates (with the one free bag policy being a brief exception). The Kingston municipal government emphasizes individual responsibility through programs such as the Remarkable Recyclers campaign (John 2012, 1), which recognizes households recycling at least 75 per cent of their waste with a special publicly displayed badge on their curbside bins (City of Kingston 2013g).

It is worth noting that the city council's deliberation and decisions about Kingston's waste future took place in the shadow of the controversy surrounding the city's Belle Park landfill, an old un-engineered dump closed in 1974 and subsequently redeveloped into a recreational facility. In 1997 a local resident took a sample of water seeping from the former landfill onto the frozen Cataraqui River and had it tested, suspecting it was killing rainbow trout in the river adjacent to the landfill (*Kingston Whig-Standard* 1997). As a result, the City of Kingston, responsible for the landfill's maintenance, was convicted of violating the Ontario Fisheries Act (*Globe and Mail* 1998), though the city subsequently successfully appealed the conviction on the grounds that landfill leachate was not an environmental threat (Tripp 2000). The Ministry of the Environment reversed the appeal in 2002, requiring the city to construct an impervious clay cap to contain the seepage (Environment Canada 2005). In the case of the Belle Park landfill contamination, the local resident and a community activist group assembled a matter of concern in a manner according to Latour's first and second forms of politics. The leachate, the frozen river, and the dead and dying fish redefined the public's understanding of the Cataraqui River (as contaminated), exceeded the City of Kingston's institutional response (denial), and generated an unsettled public. Here, politics operated according to Latour's third form, wherein the municipal government attempted to turn the issue into a clearly demarcated general will, and failed.

Leachate was likely migrating from Belle Park to the river long before it was detected due to inadequate (or nonexistent) monitoring. Objects are made visible when they are linked to other objects. Put another way, in this situation, the entanglement of waste with aquatic ecosystems, clean water, the health of future generations, recreational facilities, liberties, leachate, government, industry, and landfills rendered waste an issue. It is important here to note once again that inhuman entities such as bacteria, soil, ground water, and so on are themselves vital components of these assemblages, evincing their own proclivities to move, metabolize, and transform. Inhuman entities, in other words, play a role in transforming themselves into issues. This insight is explored further in chapter 9, where I detail the case study in Iqaluit.

As we have already outlined, because Kingston emphasizes downstream techno-scientific interventions, the city is reliant upon consulting firms that specialize in the increasingly technologically sophisticated process of siting, building, and maintaining such downstream solutions. These consultants retain their own public relations staff, engineers, and scientists, who they can use in their brokering of particular technologies to the city. In early 2007, Jacques Whitford was commissioned by the City of Kingston to begin the data collection and compilation for Phase A of the IWM study. Phase A began the process of defining the long-term WM system objectives for the City of Kingston, through looking at a needs/gap analysis of the city's waste diversion scenarios from 2008 to 2012. Several goals were identified in Phase A, including minimizing the production of residential and industrial, commercial, and institutional (ICI) waste; maximizing the environmental sustainability of the overall waste stream; and developing a system of sustainable programs and facilities "in recognition of the capability of the system to be maintained over time without exhausting the resources it needs" (Jacques Whitford 2008a, 20) and that also "minimizes costs to the taxpayer through the evaluation of overall capital investment and operations costs" (ibid., 21). From the outset, while industry collaborated with the municipal government in identifying the goal of minimizing waste production, it also explicitly advanced technological responses and cost minimization as priorities.

In December 2007, a draft of the Phase A report outlining these goals was released for public review, comment, and feedback. The report was made available on the City of Kingston website, and a public consultation session was held on 30 January 2008 (Jacques Whitford 2008a). At the session, city staff (the Environment, Infrastructure and Transportation Policies Committee) and the Jacques Whitford consultant team answered questions from the public. The city received seventy-one comments about the Phase A draft report, collected via email on the city website. The comments were compiled in the Phase A report, released in July 2008, with responses from the consultants that addressed the feasibility of the suggestions,

Figure 2.1 | Recycling collected at the Kingston Area Recycling Centre.

clarified points of concern, and answered the public's general questions about the report. In their responses to these comments, the consultants emphasized affordability, public participation, the depth of study being conducted, and – interestingly – *producers'* responsibility for waste. Jacques Whitford reiterated several times in their responses that "it is the intent of the City to consider the full range of WM approaches, systems, and technologies over the course of the study" (Jacques Whitford 2008a, 27). Yet, the upstream concerns raised by the public, such as producers' use of excess packaging, were deemed outside the remit of the local government. The response from Jacques Whitford on behalf of the city stated that "individuals/consumers likely have more influence over the use of packaging than municipalities," and such a shift would require significant changes at "provincial, federal, or international level which dictates the need for the City's system to be flexible" (ibid., 24). In this way, the responsibility to advocate for less packaging was placed on individuals and not on local government or industry itself. This is neoliberal capitalist governmentality in practical application.

Based on the goals and objectives established in Phase A, and following the initial public consultation, Jacques Whitford continued with the second phase of the IWM study in October 2008. After a background summary of Kingston's existing waste profile, the Phase B report identified a wide variety of WM approaches that the consultants claimed would improve diversion rates, including bag limit reductions, public education, reusable items diversion, waste prevention, and user-pay systems (Jacques Whitford 2008b). These approaches emphasize the responsibility of individuals to govern their waste in particular ways. The goals and objectives for Kingston's WM system were sorted into environmental and socio-economic goals that would "limit costs to the taxpayer in accordance with the environmental and socio-economic goals" (ibid., 19). The systems with the highest ranking were a residual waste processing/recovery technology in addition to the city's current approach (two-stream blue box, source-separated organics [SSO], and landfilling), and a single-stream recycling program instead of the current two-stream system. The report went on to identify eleven different system enhancement options – such as public education, additional materials recovery, and household hazardous waste diversion – that would require more research and public consultation. The most preferred options were: (1) bag limit reduction (to a one free bag policy), (2) reusable item diversion, (3) public education and system promotion, and (4) clear garbage bags.

In order to finance these options, the city added an additional $26,000 to its 2011 solid waste operating budget (Kingston City Council 2011). The one free bag policy was implemented by the council in August 2012, though it met with opposition both from within city council (the policy passed at council by a seven-to-six vote) and from the local university's student society (detailed below). The legislation was amended to provide exemptions for certain holidays and for people with certain medical conditions (provided they present required documentation), but not for households with larger numbers of residents.[2]

In 2011, the City of Kingston hired HDR to complete Phase C of the IWM study, which was to identify the residual waste processing technology that would be best suited for one of two systems identified in Phase B (HDR Corporation 2011). HDR identified four residual waste processing technologies for public consideration. These technologies included mechanical separation (the shredding of residual municipal waste to recover additional recyclable material and reduce landfilled mass), anaerobic digestion (the use of bacteria to convert organic matter in municipal waste into a combustible gas), various thermal treatments (e.g., plasma gasification, mass incineration, and waste-to-energy), and refuse-derived fuel (converting municipal waste into combustible pellets and fluff to use in power generators). HDR organized a public consultation exercise, presenting attendees (approximately twenty people, fourteen of whom were not city councillors, WM personnel, or members of our research team) with posters that included pictures and diagrams illustrating the technological processes. The public consultation, although a "key element" in HDR's phase of the study, did not motivate public support in favour of any single technology presented. Crucially, though, this event served the municipal government's requirement to provide public consultation.

After the poorly attended public consultation, HDR recommended that the city adopt either the thermal treatment option or the refuse-derived fuel option. The recommendation was based primarily on a cost analysis of the different technologies. In its report to the City of Kingston, HDR stated that "the relatively modest quantities of residual waste available in Kingston ... make the applicability of the larger scale established technologies (mass burn, refuse-derived fuel combustion and fluidized bed combustion) not economically viable" (HDR Corporation 2011, 15). HDR went on to say that the capital and operating costs associated with the other technologies, such as biological and mechanical treatment, were similarly not economically feasible due to the fact that Kingstonians were not producing enough residual solid waste. Once HDR had given its recommendation to the city, the city could continue to Phase D, which will require continued public consultation, to develop an implementation strategy for either thermal technologies or refuse-derived fuel.

Public consultation exercises hosted by the city, Jacques Whitford, and HDR have thus far garnered little public interest in Kingston, and it is not clear to what extent the adoption of a new WM technology will incite a level of public interest that might transform diverted or non-diverted waste into an issue of concern. The next time that the public may be "roused from sleep" will be in Phase D of the IWM study, when the city decides on a particular technology and identifies where the technology will be located (Jacques Whitford 2008a). However, seemingly routine deliberations among city councillors and their consultants about one technology or another may not render the issue sufficiently visible to galvanize public concern. The delayed response to the one free bag policy suggests that a public may not assemble around the issue of a new WM technology until it comes time to select a location for a new landfill or other technology, at which point the potential impact to residents may become clearer (Marres 2007).

These speculations are born out in our analysis of the media's coverage of WM in the Kingston area, including opinion pieces (in which members of the public write to the newspaper to express concerns) written by a variety of Kingston residents. Of a total of eighty-one newspaper reports we sampled between the years 2008 and 2013, from the two main newspapers serving the Kingston region, twenty-two (27 percent) were devoted to subjects outside of Kingston (such as landfill leachate emanating from the Richmond Hill landfill) that represent objects becoming issues and developing into political matters, mainly concerning government and industry culpability in creating these matters of concern, and responsibility for resolving the issues.

A much larger number (fifty-nine, or nearly 73 percent) of the newspaper reports and opinion pieces focused on waste as a matter of individual responsibility. Occasionally, the newspaper articles took a more in-depth approach, in which downstream approaches were linked to more upstream issues. The primary example of this was an article written in the *Kingston Whig-Standard*. Ostensibly about the origins of recycling in Ontario, the article goes on to explicitly identify recycling as a profit-making business, and that this business "reinforces the notion that citizens, as consumers and recyclers, are crucial to the manufacturing chain" (Schliesmann 2011a, 4). In a fascinating follow-up article, the same columnist detailed experts' concerns with recycling. The article drew particular attention to how industry tied recycling to commodity markets from the outset, establishing monetary incentives for municipalities to prioritize recycling rather than waste reduction into the system (Schliesmann 2012a). Recycling is profitable for industry and government: for instance, Kingston received $827,224 in 2010 alone from Stewardship Ontario, an organization funded by the recycling industry, for meeting its waste diversion targets primarily through recycling. As Thomas Naylor bluntly puts it, "the recycling game is a con," insofar as it sustains the consumer-based economy and does not legislate manufacturers to either reduce or eliminate packaging, or take back goods when the consumer no longer wants them (in Schliesmann 2011b).

With these limited exceptions, the major discourse found in media reports and citizen editorials fits squarely within a governmentality framework around individual responsibility, with an emphasis on self-surveillance and the surveillance of others as "good citizens." For instance, one Kingston resident wrote that "having a little enviro-guilt can be a good thing" (Switzer 2008, 1), and another wrote of a need to "curb her appetite for plastic" (Browne 2008, 1). Several opinion editorials were devoted to Kingston residents surveilling each other: neighbours and fellow residents become "bad citizens," and writers offered advice as to how citizens could transform themselves from "bad" to "good." For instance, one resident advised residents to take coffee cups home from cafes and recycle them, as she does (Toomey 2008, 1); a resident of a neighbouring township defined people who do not want to use clear bags for recycling as "anti-recycling," people who want to "hide (stuff)," and "not true recyclers" (Jefferson, in Edmiston 2010, 1). Concomitant with this internalization of responsibility for waste and surveillance of others is a familiar discourse on tax fairness. For instance, one opinion editorial writer wrote that "producers [and] consumers already pay a fair share" for waste disposal (Sonnenberg 2011, 1). City councillors were very aware of this concern. As councillor Bryan Paterson expressed after the one free bag limit was introduced, "I think it's great, but this is reducing a service without a corresponding decrease in property tax. I think the optics of it are terrible" (in Norris 2011, 2).

A framework of neoliberal governmentality helps make sense of the fact that the closure of Kingston's landfill to the public in 2011 did not garner much public interest. Instead, residents' attention was focused on the approval of the one free bag policy, because the policy involved individual responsibility, as well as the high value that neoliberal governance places on what it calls freedom of choice (i.e., the freedom of consumers to consume what and how much they like). Having remained "asleep" during the initial public consultation exercises, members of the public – particularly the university's Alma Mater Society and the Kingston Rental Property Owners Group – "woke up" only after the policy was implemented.

These groups made speeches to city council, stating that the policy would be inequitable for larger families, would not provide a meaningful contribution to diversion methods, and was unrealistic for university students who live off-campus in multiple dwelling homes (Alma Mater Society 2012; VandenBrink 2012). Troy Sherman, the municipal affairs commissioner for the university's Alma Mater Society, argued that it would "force students to pay out-of-pocket for additional garbage bags," leading to student "cynicism," and that it was "unrealistic" for students to coordinate the efforts of students living in a single dwelling: "This policy demands collective action from a household, which can be next to impossible in a student environment," Sherman stated (in VandenBrink 2012). This rather remarkable statement would surely be contested by the hundreds of thousands of university students worldwide who coordinate, for instance, to protest various forms of injustice in Quebec.

As Marres (2007) points out, diversion rates, bag limits, and property taxes became public issues worthy of civic involvement only when these things were perceived as a significant change to people's lifestyles and/or livelihoods. Here, governmentality is expressed as a combination of neoliberal assertions of individual freedom, such as Queen's University student representatives arguing that they should be able to dispose of as much garbage as desired without paying more, and the responsibility of the government (to dispose of waste but not increase taxes). This said, the one free bag issue quickly went back to sleep as media coverage increasingly took on a positive tone, and emphasis was placed on individuals governing themselves through WM surveillance practices (MacAlpine 2012).

## MAKING WASTE MANAGEMENT PUBLIC

This research suggests that Kingston's experience with WM is by no means unique in Canada. Municipalities across the country are grappling with what to do with non-diverted waste, and indeed, with increasing amounts of diverted waste. There are several reasons why interest in waste nevertheless tends to be relatively short-lived, and therefore why waste as an object only occasionally transforms into an issue that generates political action. Indeed, the ubiquity of landfilling in Canada offers numerous examples of objects associated with waste made into issues, then assembling publics, becoming political, and then largely falling back to sleep. For instance, we might contrast the politics in Kingston with those in Simcoe County somewhat earlier. Scott Lougheed found that in 1986, Simcoe County concluded a search to locate a new landfill, choosing a twenty-one hectare parcel of land in Tiny Township called Site 41. Site 41 is situated atop the Alliston aquifer, said to have water of exceptional purity (Shotyk et al. 2005; Shotyk and Krachler 2009). Despite over a decade of ongoing public opposition to the landfill, the Ministry of the Environment granted a Certificate of Approval in 1998 to construct a landfill, pending approval of a feasibility report.[3] This feasibility report represented a primary technique used to render the landfill a non-issue and therefore prevent the assembling of a political response around it; it limited the parameters of debate to the design and operation of the prospective landfill, rather than the long-term environmental and health impacts of the landfill, and the myriad upstream issues produced by tying economic growth to waste production (Schnaiberg 1980).

Needing to demonstrate its accountability to the public, the ministry permitted the formation of a Community Monitoring Committee (CMC), comprised of residents from the surrounding area "charged with the responsibility of providing community review of the development, operation, on-going monitoring, closure and post-closure care" (Ontario Ministry of the Environment 1998, 11). By assembling around such objects as aquifers, leachate, and hydrological models, the public slowed down the county's attempts at a swift, formulaic process, described by Latour's third sense of politics. Turning objects (for instance, leachate in soil) into issues (the environmental contamination of an underlying aquifer), members of the public constructed a politics around "knowledge controversies" by commissioning their own peer reviews of the siting and design documents, engaging an independent landfill engineer, and gathering their own scientific evidence. This enabled them to question the county consultant's emphasis on the certainty of short-term risks. For example, Kerry Rowe, one of my research collaborators, noted that two years of data were omitted from the consultant's hydrological models, and long-term outlooks on the effect of climate change and water use were not considered (Millar 2008, 7; Rowe 2004, 1–2).

County residents transformed waste into an issue using existing institutional procedures and by entangling physical-technical processes (leachate, hydrology, water purity) with social processes of risk definition and democracy (risk thresholds of permissible human contamination, and community group activism) (Marres 2007, 770; Whatmore and Landström 2011, 2). Outside the county hall meeting at which plans for Site 41 development were finally voted against, a brief moment of celebration was swiftly followed by the public's disassembling and dispersal (Friesen 2009; Latour 2007b). After a cost of over $11 million and nearly twenty-five years of protest and negotiations, county residents will inevitably face waste as an issue again, as increasing waste export costs will lead the county to consider other WM options, including new landfill siting. This means that while Simcoe County's public has, for the time being, fallen back to sleep on this issue, there may be further occasions for government and industry to make new claims about – and politicize – WM processes.

## ASSEMBLING POLITICS

Latour's fifth sense of politics takes up Foucault's theory of governmentality wherein certain objects are encountered in such a ritualized fashion as to not garner the attention required for them to be transformed into issues. Applying Foucault's theory, these are the practices through which individuals govern themselves, internalizing the erroneous assumption that most waste is produced post-consumption and that waste diversion and disposal constitutes "good citizenship." Governmentality structures the continued conceptualization of waste as an object over which individuals govern themselves, and thus largely obviates the transformation of waste into a matter of political concern demanding critical attention. In other words, Latour's fifth sense of politics points to ways in which waste is a mundane object unworthy of sustained public action. This is a situation easily entered into but difficult to get out of (see also Downs 1972). As we have seen, decision-making about waste in Kingston has been a largely bureaucratized process consisting of formulaic consultation with WM industry, a city staff report for councillors to consider, and a small number of town hall meetings to provide a structured (and predetermined) means of citizen participation. The staff report is based upon the WM industry consultation, which recommends particular WM technologies based on current and projected amounts of waste that the city creates, and the cost of various WM technologies. On the few occasions when members of the public expressed upstream concerns such as having opportunities to be more involved in decision-making regarding the future adoption of another WM technology, these expressions were deemed to be outside the remit of municipal government and industry, and overshadowed by the need to maintain, and indeed, increase individual responsibilization and neoliberal concerns.

Our study suggests a close association between Latour's fifth and third political forms. When waste becomes denaturalized and non-routine, and begins to transform into an issue, Latour's third sense of politics frequently emerges. The perennial concern with increasing volumes of waste is, as we have shown, usually met with a combination of individual-level surveillance, purported to decrease waste volume, and techno-scientific fixes such as landfills, incinerators, and waste-to-energy technologies. These fixes are presented to the public in such a way as to suggest that the issue has already been resolved and simply requires public accession (Spector and Kitsuse 1972). For instance, by the time Simcoe County began public consultations, the county had already established the need for a landfill, selected the site, identified the risks and assessed them to be within pre-determined acceptable values, and chosen a developer. The increasing hegemony of multinational WM corporations – corporations adept at managing public involvement – means that industry increasingly defines the parameters of assessment and ipso facto what may be examined, discussed, deliberated, and agreed upon. The industry-produced WM feasibility studies that increasingly inform government deliberations and decisions are focused on economic profitability and are supported by in-house science and engineering reports that emphasize techno-scientific responses with known short-term risks. Multinational corporations have become vital allies through which municipalities attempt to turn an emerging issue into a technical problem governed by a clearly demarcated general will and common good.

Members of the public participate in these political deliberations at a considerable disadvantage, on their own time and with modest or no funding. Recently assembled and assembling groups must galvanize evidence and arguments against for-profit companies that have scientists, engineers, contractors, policy specialists, and experts in old and new media communication on payroll (Hannigan 1995). This said, Latour's third sense of politics pivots on the failure of governments to foreclose public dissent. As our analysis suggests, issues that are apparently already resolved may be complicated by a skeptical public focused on long-term risk and scientific indeterminacy, and who may raise more upstream issues such as de-growth, circular economy actions, zero waste, and/or producer responsibility goals. As we have seen, members of the public are wary of scientifically described risks – which are, by definition, known risks typically involving numerical thresholds of *a priori* defined acceptable environmental compromise – and focus as well on unpredicted effects, which scientists cannot address (Wynne 2006, 216).

Site 41 and Kingston's Belle Park demonstrate the type of dissent that can lead to the failure of government to present WM as an already-resolved issue: power is neither unidirectional nor located in only one entity, such as industry. These politics have, thus far, largely oscillated between Latour's fifth and third forms, and further comparative analysis with other municipalities may provide insights into how other forms of politics take shape. This is all the more interesting because relentless circuits of production and consumption coupled with an almost exclusive focus on techno-scientific amelioration ensure that increasing waste remains an unstable object, primed to be transformed into an issue.

## UPSTREAMING WASTE ISSUES

While the selection and siting of a waste management technology such as an EfW facility within the city will affect residents, what is considered effective public participation by city staff is rather limited. For example, the single information session specifically addressing the selection of a local waste management technology[4] was attended by twenty-two people, three of whom were members of our research team, representing a fraction of 1 per cent of Kingston's voting population. This was described by city staff as "a good turnout, considering the waste processing planning for Kingston is still in the early stages" (Stafford 2011). The implication is that public involvement is heavily circumscribed and should only take place once the issue and solution have been established, rather than when the issue is being defined. Consistent with past studies (Chapman 2004; Darier 1996a, 1996b; Hobson 2013; Petts 1997), the majority of our respondents expressed the desire to participate in WM decision-making, and most respondents reported that the public should be heavily involved. However, only a fraction of respondents were aware of opportunities to participate, and our observation of public consultation events suggests that even fewer actually did participate.

Additionally, respondents were generally unaware that a private industry consultant had heavily "guided" the city's deliberations.

Thus, neoliberal governmentality generally sets the parameters of participation in political deliberations, and governments tend to prefer that such matters remain non-issues for the public. The public is given a limited set of waste management options, selected by HDR in a political process that the public did not participate in and of which the public remains largely unaware. As Petts (1997) observes, what municipal governments and industry term "public consultations" may be better understood as public relations exercises designed to exhibit information and facilitate community assent, rather than a process that encourages members of the public to raise concerns. At the information session referred to above, members of the public were presented with five options, chosen by the private industry consultant HDR as potentially suitable for the city. Documentation provided to attendees strongly emphasized thermal treatment methods as the most suitable choice (HDR Corporation 2011). Data collected from attendees during the session by HDR suggested a high degree of support for a thermal facility (ibid.), while our results indicate that among the minority who did select a technology, mechanical separation was favoured. This discrepancy is not surprising, for as Hobson notes, outcomes such as support for thermal processing may be "the consequence of entering into the complex field of contingent governance" such as public consultations, "the momentum of which is not ... sustained beyond the life of the intervention" (2013, 68).

Consultations are an opportunity for municipalities (and thus the WM industry) to address the putatively erroneous or deficient knowledge of the small number of residents concerned enough to attend, and garner their assent. This exemplifies the "paternalistic" tendency of local governments and hired experts to provide "the 'right' answers" (Petts 1997, 361). Thus, in the Kingston case study, rather than the consultation process assessing residents' opinions after a critical and open discussion, the support that the process obtained may itself be seen as a product of the consultation context and the specific relations of power involved. Thus, more meetings and more consultation may not be an adequate response and may well perpetuate the prevailing neoliberal governmentality as a technology "of transformation, meant to achieve and to demonstrate the educational effect of carefully conducted citizen deliberation" (Braun and Schultz 2010, 410).

But people are not mere "blank slates" passively receiving instructions (Darier 1996a, 591), nor are they subject to a single governing project (Malpas and Wickham 1995). Competing subjectivities may, for example, dictate that producing less waste to begin with is valued more than recycling, or that recycling is not worthwhile to begin with. It is precisely because of this resistance that projects for governing household waste are sustained and intensified, resulting in additional efforts to normalize these projects, such as increased education programs and the Remarkable Recyclers program (which has received approximately fifty participants in its largest year). Further, residents are evidently uncomfortable with the broader neoliberal tendency to involve industry in governance. This is indicated by the fact that residents rated industry and industry scientists as the least desirable stakeholders in WM decisions, despite them being the most heavily involved (Lougheed, Hird, and Rowe 2016). And yet, residents were not provided an opportunity for involvement until after the local government had partnered with industry, thus precluding any possibility of resistance to this particular institutional arrangement.

CHAPTER THREE

# Looking for Redemption in All the Wrong Places: Hiding Our Waste and the Cult of Recycling

## INTRODUCTION

As the previous chapter argued, issues over waste garner relatively little public attention or democratic dialogue: the circuits of capital – production, consumption, and waste – means there is essentially "no appeal to a citizen who is not also a consumer" (Easterling 2014, 208). In other words, Canadians' identity as consumers is amplified, to the diminishment in equal measure of our identity as citizens entitled to – and indeed responsible for – making demands of our government – in this case that government (at all levels) address waste as an upstream issue (i.e., focusing on waste reduction and industry responsibility).

If waste bears witness to all those things we have discarded and with which we no longer want to be associated, then the highly lucrative waste industry is happy to oblige our desire to remove the traces of our consumption. An increasingly vital part of the aftercare of a landfill is its removal from memory: landfills are covered over with grass, family parks, or suburban housing, or made into ski hills in an attempt to repurpose the land and increase profit, while masking the waste buried beneath. When brought (back) into view, waste landscapes are inevitably the "material sedimentation of destruction" (Gordillo 2014, 119), with imprescriptibly devastating consequences for present and future generations, as communities such as Love Canal discovered at considerable cost.[1]

The waste we do not store in perpetuity underground is transported to other regions, again almost entirely out of sight and out of mind. There is, as Kathryn Wheeler and Miriam Glucksmann point out, a strong "moral economy" that operates such that the "responsible 'citizen-consumer' is motivated to act because of his or her commitment to moral/political projects rather than in line with his or her selfish desires" (2015, 143–4). Canada is not alone in adopting a moral economy to recycling as emblematic of the "good environmental citizen": studies show that other countries such as the UK and Sweden have adopted this discourse wholesale (ibid.; Skill 2008). Indeed, Skill (2008) demonstrates that around the globe, "recycling is the most common action that households regularly performed" (in Wheeler and Glucksmann 2015, 153). In the UK, this moral economy extends to saving public money through recycling. And in Sweden, Swedish-raised respondents distinguished themselves from immigrants as the "irresponsible other," based on the perceived lack of proper recycling by immigrants (Sayer 2005). This chapter takes a closer look at how and why recycling has become the most common response to Canada's waste problems, and emblematic of caring for the environment. Together with Cassandra Kuyvenhoven, whose PhD research concerns the environmental impact of transporting waste and recycling, I wanted to focus more closely on the issue of moving waste around, and its impact on mitigating the claims that recycling is "better" for the environment than landfilling. Cassandra and I continued to use Kingston as a case study because Kingston is a mid-sized city in southern Canada with similar challenges to other southern Canadian cities, such as a closed landfill, which necessitates the transport of 100 per cent of the city's waste and recycling to other regions.

## MODES OF WASTE GOVERNANCE

The "waste hierarchy" organizes the management of waste from most preferable (reduce), to the less preferable (reuse), to the even less preferable (recycle), and finally to the least preferable (disposal) (Gharfalkar et al. 2015). The European Commission (2008), for instance, states that the waste hierarchy provides a good measurement of a region's modes of waste governance (Hultman and Corvellec 2012; Kurdve et al. 2015; Ewijk and Stegemann 2014).

The creation of the waste hierarchy and its implementation has a lot to do with what we might think of as the changing "identity" of waste, which pivots on a region's shifting political, economic, and cultural demands. A number of scholars have examined critical

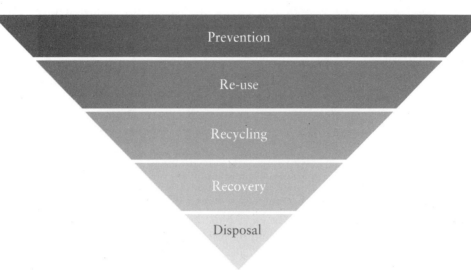

Figure 3.1 | The waste hierarchy, a hierarchy of options for managing waste in terms of what is best for the environment.

shifts in how waste is perceived, and as a result, how it is managed. Bulkeley, Watson, Hudson, and Weaver (2005) offer perhaps the most detailed analysis of the development of the waste hierarchy, coincident with changing political priorities and constraints. These researchers identify four coexisting "modes of governance" – disposal, diversion, eco-efficiency, and waste as a resource – that correspond to the waste hierarchy (ibid.). Certainly, these ways of governing waste overlap and inform each other, but they nevertheless reflect a general organizational (if not paradigmatic) shift from disposal (the bottom end of the waste hierarchy), to diversion, to waste as a resource to be repurposed, and finally to the ultimate goal of reducing waste. The articulation of the most and least preferable means of dealing with waste is unsurprising within the context of global neoliberal capitalism. Zsuzsa Gille (2007), for instance, identifies a paradigm shift in Hungary's conceptualization of waste from pre-war socialism (which emphasized reuse and recycling) to post–Cold War capitalism (which emphasized disposal).

Canada (as elsewhere) has come to heavily emphasize the lower end of the waste hierarchy (disposal and diversion) at the expense of the higher end (reduction and reuse). Indeed, diversion – in the specific form of recycling – has come to dominate the public's interactions with the waste they are most familiar with, that is, municipal solid (household) waste. And in practice, "diversion" has become a stand-in for "recycling." That is, the waste hierarchy has become so entrenched that the details of both its assumed priorities and its actual efficacy have taken backstage to what is now a rather frenzied and noisy focus on recycling and the "individual responsibilisation" of waste (see Darier 1999; Hargreaves 2012; and Oels 2006). In other words, the details of the management of waste often overwhelm or underwhelm (depending on your perspective) the public imagination. So swayed are we by the economic, environmental, and moral merits of the diversion of waste through recycling that these priorities have become the mantra of waste. In the rush to appear to be more environmentally conscientious, municipalities across Canada (and the globe) are increasingly emphasizing recycling.

## The Disposal Mode

A number of scholars document historical shifts in the governance of waste in North America from individual cartage and disposal operators to the formation of much larger industries and corporations such as Progressive Waste and Waste Management (Crooks 1993; Davies 2008; Melosi 2005). The disposal mode of governing coincides with this shift, as individual cartage operators seeking opportunities for the reuse and recycling of discarded materials gave way to burgeoning WM industries working in tandem with local governments to remove and dispose of various forms of waste, including municipal solid and liquid waste and some forms of industrial, commercial, and institutional (ICI) waste (see chapter 2). These WM industries secure ever-increasing profits by focusing on service delivery (i.e., removing waste from sight) and economic efficiency (which only includes recycling materials that are profitable, and disposing of the rest) (Bulkeley, Watson, and Hudson 2007, 2,741). Without question, landfilling is the preferred mode of waste disposal in Canada, and open dumps have given way to modern engineered landfills in southern parts of Canada. As Bulkeley et al. succinctly summarize: "with holes to fill and rubbish to get rid of, landfill has made

Figure 3.2 | Park garbage can where diligent park users have deposited bags containing dog waste as close to the overfilled garbage can as possible.

almost incontrovertible (economic) sense as a waste management option in recent decades" (2005, 3).

As the previous chapter described, within the disposal mode, waste is primarily governed through two relationships: the contracts local municipalities establish with WM industries for waste removal and disposal (which indirectly includes all of the cascading relationships WM industries make with other companies for the disposal and diversion of the waste), and local governments' relationship with their constituents, which emphasizes both municipal solid waste (MSW) and individual responsibility for this particular form of waste. The disposal mode is clearly a downstream response to waste. Emphasis is diverted away from upstream issues, such that the much greater volumes of waste produced by industry and the military, and instead directed towards techno-science "fixes," such as the merits of incineration versus landfilling, and more technically robust longer-life landfills. As Maniates notes, "confronting the [production and] consumption problem demands, after all, the sort of institutional thinking that the individualization of responsibility patently undermines" (2002, 46–7).

As the amount and diversity of waste has steadily and at times dramatically increased in the 1980s and 1990s, as landfills began to reach or surpass capacity in the 1980s, and as the environmental impact of waste became a focus for the growing environmental movements, a certain amount of scepticism towards unending disposal set in. According to Bulkeley, Watson, and Hudson: "as environmental issues moved up public and political agendas, the regulation of waste disposal became subject to tighter scrutiny" (2007, 2,741). Grassroots organizations, municipal governments, and the WM industry itself began to refer to the waste hierarchy and a need to emphasize diversion.

*The Diversion Mode*

The Waste Diversion Act, introduced in Ontario in 2002, for instance, sought to promote "activities to reduce, reuse, and recycle the designated waste" and "not to promote ... the burning of designated waste, the landfilling of designated waste, or the application of the designated waste to land" (Government of Ontario 2002, 8). Local governments and industry primarily endorsed and sponsored – almost exclusively – the implementation of recycling programs (known as blue box recycling). In the early 2000s, the Ontario provincial government set an ambitious diversion target of 60 per cent, to be reached by 2008.[2] This was revised by the Ministry of the Environment, Conservation and Parks in 2019 to diversion targets of 30 per cent by 2020, 50 per cent by 2030, and 80 per cent by 2050. As such, as Bulkeley et al. point out, "the achievement of targets, in this case for recycling, composting, and the diversion of biodegradable waste from landfill, becomes the basis of governing, so that the conduct of government is self-reflexive" (2007, 2,747).

Thus, within the diversion mode of waste governance, diversion targets themselves become the objects to be governed at the municipal level: "the original governmental rationale of a concern with the global environmental implications of waste management is deflected to the narrower preoccupation with the diversion of waste from landfill" (ibid.). Diversion has certainly not replaced disposal, and the latter remains the primary (if now less talked about and more discreet) mode of waste governance. Yet diversion is very popular with local governments because it places the emphasis on MSW (as opposed to the far greater quantity of industrial waste) and individual responsibility. Diversion is also very popular with WM industries, which often own and control the recycling companies. And this is absolutely key to consider: individual citizens and environmental organizations have enthusiastically clung to recycling as a primary solution to waste generation, despite the fact that whether a municipality's waste is disposed in industry-owned landfills or industry-owned recycling companies, industry profits. Indeed, recycling has become emblematic of caring for the environment in general rather than a WM technique (Liboiron 2010).

Diversion is presented as a societal and environmental "good," but the reality of diversion is more complicated, and certainly not a complete solution to waste issues. Recycling is a downstream and short-term response, which does nothing to stem the tsunami of ever-increasing production and consumption. Recycling materials requires that they be reprocessed, and this requires a great deal of energy, often using non-renewable fossil fuels that pollute the soil and the atmosphere. Recycling may also release hazardous wastes through by-product emissions, and it produces lesser-quality products that may only be used once or twice more (Center for Sustainability 2012; MacBride 2012; Rowe 2012). Recycling paper requires the significant use of toxic chemicals (such as chlorine) to remove ink and generates its own waste – sludge – that is more difficult to dispose of than paper (United States Department of Energy 2006). Using paper products requires harvesting wood, which in turn requires:

> manufactured heavy equipment and vast quantities of petroleum products. Composite wood products contain glues made from petroleum ... when wood is made into paper, only part of the wood, the cellulose, is used. The other part, the lignin, becomes waste ... De-inking and re-pulping include a mechanical treatment. This damages the cellulose fibers. They break and the average fiber length decreases. Fiber length has a significant effect on the quality of the paper, therefore the fiber cannot be used indefinitely. (Baarschers 1996, 181, 190)

Indeed, removing ink from the paper in the recycling process inevitably causes a loss of between 10 and 20 per cent of the fibre (ibid.). In other words, each time paper is recycled, it produces a degraded product, and it cannot be recycled indefinitely.

Moreover, the "dirty little secret" of recycling is that, due to constantly varying markets, materials the public believes are destined for diversion are in fact landfilled or incinerated: waste intended for recycling is often disposed of when recycling costs outweigh the profit derived from the recycled materials, because the same industries own both recycling processing centres and disposal centres. Moreover, evidence suggests that people actually produce more waste when they have access to recycling (S. Harris 2015). Moreover, Max Liboiron makes the important point that industry likes recycling programs because it externalizes the costs to consumers:

> The Container Corporation of America sponsored the creating of the recycling symbol for the first Earth Day in 1970. The American Chemistry Council, the world's largest plastics lobby, enthusiastically testified in favour of expanding New York City's curbside recycling program to accept rigid plastics. Recycling is a far greater benefit to industry than to the environment ... Industry champions recycling because if a company has reusable bottles, for example, it has to pay for those bottles to return, but it if makes cheap disposables, municipalities pick up the bill for running them to the landfill or recycling station. (2013, 10–11)

Finally, we need to consider another crucial aspect of the material limits of recycling. For example, even if we were somehow able to recycle all of the metals and minerals contained in electronics such as laptops, smart phones, and cars, the current demand for electronics has already outpaced the supply that recycled electronic waste could produce. As figure 3.3 shows, as electronics demand increases, the minimum gap to be filled by primary resource extraction simply increases as well.

The benefits of recycling are particularly compromised when we consider one aspect of recycling that typically receives little attention. Transporting waste to reach recycling processing facilities significantly mitigates claims that recycling is always the environmentally better option:

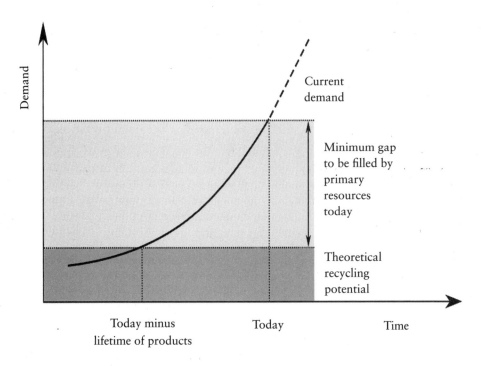

Figure 3.3 | The growing gap between available metals and minerals, and current and growing demand.

> Recycling requires collection, transportation and, at the recycling plant, industrial-scale processing. Transportation and processing mean energy consumption as well as processing wastes. The recycling industry is an industry and it inevitably produces its own industrial waste. (Baarschers 1996, 187)

As well, materials often need to travel to more than one reprocessing centre in order to be recycled, again requiring more transportation than directly disposing of waste. And then, whatever waste is recycled will eventually (typically after only one reuse) turn into waste that will be transported to a landfill (or less often, an EfW facility or incinerator). And whether waste is disposed of or recycled, it must inevitably be transported, with the attendant problems of transportation, including the very significant use of non-renewable fossil fuels, truck carbon emissions, road creation and maintenance issues (including the increased use of materials and pollution), the increased production of waste transportation vehicles, the increased risk of road accidents, and so on. As William Baarschers notes, "after the diesel fuel is burned, the carbon dioxide and nitric oxides produced, we have waste tires, waste batteries, and waste trucks at the end of the line" (1996, 190). Moreover, in an attempt to meet diversion targets, many municipalities have resorted to the transportation of waste to other municipalities, provinces, and/or countries – to be disposed of somewhere where it will not affect their provincially directed diversion targets. Transporting waste longer distances only increases the use of non-renewable fossil fuels and the other problems just noted. Taking these myriad considerations together, recycling is far more complex, and far less of a "societal good" than it appears on the surface. Later in this chapter, we will take a look at this issue in more detail, with our case study from Kingston, Ontario.

## The Eco-Efficiency Mode

Given these realities, the model used by Bulkeley et al. examines two further modes of governance. The authors argue that a third, emerging eco-efficiency mode attempts to "[move] waste management options up the waste hierarchy" (2007, 2,748) towards reuse rather than diversion (i.e., recycling) and disposal. Indeed, this third mode prioritizes the reduction of negative impacts of waste and WM through "more dramatic and transformative action, including an emphasis on material reuse instead of more energy-intensive recycling" (Pollans 2017, 2,303). To achieve this, the eco-efficiency mode is characterized by partnerships with state and non-state stakeholders (and, we would expect, rights holders, although Bulkeley et al. do not refer to rights holders specifically) in an effort to extend governmental authority, while enrolling "specialist expertise" offered by private industry (Bulkeley, Watson, and Hudson 2007, 2,748). A central aspect of the eco-efficiency mode is the deployment of different forms of governmental technologies, "including those which use 'inducement,' such as financial incentives, as well as those, such as joint working within the authority and with other partners" (ibid.). Bulkeley et al. contend that, in some places, "this mode of governing is creatively redefining the rationality of municipal waste management away from focusing on targets to achieving more concrete, local, social, and environmental objectives" (ibid.).

## Waste-as-Resource Mode

The fourth and final waste-as-resource mode of governance attempts to reframe objects that are conventionally seen as waste into resources that may or may not be economically profitable (Hawkins 2006); the emphasis is on "social and environmental" benefits through the reframing of waste (Pollans 2017, 2,303). This emphasis corresponds well with closed-loop or circular economy discourses, which stress the continuous circulation of materials within the economy. As such, this fourth mode of governance resembles the other three modes insofar as it emphasizes MSW as any municipality's greatest waste problem, and individual responsibility for waste as any municipality's greatest waste solution. Indeed, this mode of governance attempts to promote MSW (and it is uniquely MSW, and not the greater volumes of industrial waste a municipality produces) as a conduit for organizing communities around shared values and commitments: "waste is not only a material resource, but also one which provides opportunities for skills development and employment and an entity to be governed through the mobilization of communities" (ibid., 2,749). A waste-as-resource approach values economic resources (i.e., when waste is transformed into energy via technologies), but also prizes environmental and social resources (i.e., community-building and positive changes to the environmental landscape). As Pollans notes, other environmental and social benefits may include:

> the consideration of good, well-compensated jobs generated through the repurposing of goods or reprocessing of materials, as well as the ecological benefits from reducing the extraction of virgin materials, reduced disposal, the return of nutrients to soil, and reduced transport of waste and virgin materials. (2017, 2,303)

Insofar as waste-as-resource decreases the amount of waste (there is no inherent incentive to decrease the proliferating *kinds* of waste humans produce), it is difficult to argue against this mode's proximity to the apex of the waste hierarchy (reuse and reduction). At the same time, the primary means of achieving this – defining waste as "resources out of place" (to twist Mary Douglas's aphorism [2007]) – is by identifying the public as a primary conduit of certain micro-processes of neoliberal capital accumulation that work to effectively mobilize discourses of environmental stewardship, which can often be used to rationalize such things as charity shops, furniture reuse, and community composting projects. As Pollans notes, waste-as-resource "redefine[s] waste management as a process of social and environmental stewardship" (2017, 2303). As is emphasized by the discourse of social responsibility for the management of waste, individual-scale responses increase as we move from first to fourth modes of governance (and indeed, as we correspondingly move up the waste hierarchy), the fourth mode ironically represents the most robust entanglement of publics with neoliberal capitalist accumulation, while dissuading the consideration of waste as a socio-ethical issue. If, as Pollans argues, the fourth is an "aspirational" mode that has yet to be attained in North American societies, we should carefully consider the connections between modes of waste governance, neoliberal capitalism, and public responsibilities – a sobering consideration I return to later in this chapter.

## REDEMPTION IN PRACTICE

As we have seen, Canadian waste is governed at the federal, provincial/territorial, and municipal levels. Federal-level laws mainly concern waste issues that supersede provincial/territorial or national boundaries (Parliament of Canada 2003). Provincial and territorial legislatures assert their independence from the federal government and power over municipal governments in regard to governing waste disposal sites. Municipalities are responsible for implementing their individual WM systems. While abiding by the rules of the province or territory (a flow of governance which is consistent across Canada), municipal governments are responsible for planning and proposing WM systems, which includes a final waste disposal site. However, municipalities have the option to outsource WM systems, in entirety or in part, to industry. As such, the day-to-day management of, and future planning for, waste is largely governed at the municipal level by municipalities in close relation with industry. In practice, this means that Canadian municipalities are moving their waste in greater amounts. For instance, between 2012 and 2016, Ontario's waste export to Michigan increased by 66 per cent to over 3.2 million tons (OWMA 2016).

In order to more clearly examine how recycling has become "the solution" to our waste problems, I return to the case study of Kingston, Ontario. As we learned in chapter 2, since the management of MSW has been privatized in Canada, Kingston's city council enters into contracts with WM companies to perform waste disposal activities for the municipality, based on a request for proposal (RFP) that is prepared by the city council's Solid Waste Division, which is part of the city's Public Works Services Department. The Solid Waste Division serves the Kingston, Loyalist Township, and South Frontenac areas and is responsible for operating and maintaining the city's solid waste services (including garbage collection, the green bin program, and recyclables at curbside), as well as managing the Kingston Area Recycling Centre (KARC), the material recovery facility, the household hazardous waste facility, and the composting and brush sites (City of Kingston 2016). The employees of the Solid Waste Division are hired, not elected, and are required to have their budget and initiatives approved by the city council. And according to the local waste managers employed by city council that we interviewed in our larger study (see chapter 2), the council usually accepts the Solid Waste Division's recommendations, unless there are budget issues.

The RFP includes several criteria that are used to evaluate WM industry proposals, based on four criteria. The first criterion is "Pricing and Related Costs," which accounts for seventy-five points out of one hundred (City of Kingston 2014, 3). The other evaluation criteria include "Company Profile and References" (twelve points), "Proposed Methodology" (ten points), and "Accessibility Standards for Customer Service" (three points) (City of Kingston 2014). As such, the most significant criterion is, explicitly, *service cost*. Each bid must include the projected annual cost for waste services, based on the city's estimated annual tonnage of waste.[3] According to our respondents, the system awards the most points to the applicant with the lowest bid. This makes it difficult for smaller, more local waste companies to make competitive bids against larger WM industries like Progressive Waste or Waste Management of Canada because larger companies and corporations own more (if not all) of the stages of waste disposal (waste removal trucks, the landfill, the recycling processing plants, and so on).[4] Because Kingston no longer has an open landfill and has no recycling facilities, the waste contractors must have access to facilities outside of the municipality to dispose of and divert waste. Put plainly, because pricing makes up 75 per cent of the evaluation process, larger WM industries can often significantly underbid smaller companies. And larger WM companies are far more likely to own and operate both landfills and reprocessing (recycling) centres, and thus control whether waste is disposed of or recycled, according to which is most profitable.

According to a former manager of KARC, environmental considerations – including, for example, the distance that waste may be transported for disposal or diversion/recycling – would technically be included under the Proposed Methodology section of the RFP. However, when asked about the likelihood of including stricter, more upstream environmental controls (i.e., local disposal only, an eco-friendly fleet of trucks, etc.) in the RFP, one respondent in the Solid Waste Division said that the RFPs include as few restrictions as possible so as not to alienate potential bidders. And alienating potential bidders is inadvisable because Kingston is entirely reliant on industry to handle the municipality's waste.

## Disposal

The most recent RFP for collection and disposal services had only two "valid" applications, which fell short of the requisite "three valid responses" as outlined in By-Law Number 2000-134 (City of Kingston 2014, 61).[5] Of the two valid responses, Waste Management of Canada received the highest score, due in large part to being $51,300 less than the second proposal, made by Tomlinson Environmental Services, a Kingston-area WM company. Once under contract, the industry becomes responsible for Kingston's waste (i.e., disposal, diversion, and conversion into resource) (Ministry of Environment 2015).

From here, the network of actors involved increases, as the contracted WM industry is able, at its discretion, to subcontract any part of the waste removal or processing. With no open landfill within city limits, disposal of Kingston's waste within the municipality is not a possibility; therefore, the waste needs to travel at least beyond the city's borders (and often much further than that). So, the initially contracted company might report back to the municipality that Kingston's waste ended up at a transfer station in Moose Creek, Ontario, but they are not responsible for reporting what happens to the waste after it arrives at Moose Creek.[6] Thus, while municipalities are entering into agreements with industries in order to meet their

WM needs, they are also tangentially entering into the relationships that industries build with other waste subcontractors and/or reprocessing centres. These multiple relationships create a constellation of non-state actors (primarily industries) that are making decisions about where to dispose of Kingston's waste, based on economic (profit) concerns. This is how Vancouver's waste ended up in the Philippines. And as such, Kingston residents – and municipal officials – are largely unaware of where Kingston's waste travels. When asked about Kingston's practice of transporting waste outside of the city, none of the city councillors interviewed knew where Kingston's waste was disposed, and, consequently, were uninformed about the distances that Kingston's waste travelled. The councillors were more concerned with other waste issues, like achieving the waste diversion targets set out by the city, littering, and ensuring that Kingston citizens were properly educated on how to properly sort their waste into the two-stream system (disposal and recycling).

Therefore, although city council does not explicitly prioritize disposal in its waste hierarchy – the municipality maintains that it is focused on achieving waste diversion (recycling) targets – disposal remains the most prominent mode of governance within the municipality through practices such as prioritizing the lowest contract bid. The concern that city councillors will lose their seats if taxes for waste disposal and/or diversion increase may well translate into an implicit reluctance to include more eco-friendly initiatives in its RFPs.

*Diversion*

Following the release of an integrated waste management (IWM) study in 2010, the City of Kingston announced its goal of becoming "Canada's most sustainable city" (City of Kingston 2016). In an effort to become more sustainable in their waste sector, city council announced that it had achieved a 60 per cent diversion rate by 2015, which has now flatlined (City of Kingston 2019). When discussing the waste issues Kingston faces, several city councillors emphasized that Kingston was struggling to meet its targeted waste diversion rate. According to a city councillor interviewed by the local newspaper:

> You can do it [recycling] voluntarily or governments are going to make you do it. We made really good strides to begin with, but eventually the low-hanging fruit is not there. It's mostly a

cultural issue of persuading people this is really necessary and we need to do it. (In Schliesmann 2015, 1)

Emphasizing the responsibility of its citizens, the City of Kingston has adopted several initiatives that explicitly emphasize diversion in the form of recycling. Over several years, the city introduced a diversion system using grey bins (paper), blue bins (glass and plastic), and green bins (composting), which were given to each household. In 2012, it also introduced an initiative to reduce the amount of waste through the adoption of a one free garbage bag policy (MacAlpine 2012). Given the evidence that consumers tend to increase their consumption when they know there is a recycling program in place (Catlin and Wang 2013; Sun and Trudel 2017), it is unclear to what extent the further restriction to the number of bags of garbage may be increasing the amount of waste designated for recycling. City councillors unanimously stated that recycling is a very significant part of Kingston's WM system, and, when interviewed, several councillors stated that they hoped to reach a diversion rate of 100 per cent. Thus, government officials do not look beyond the uncritical assumption that recycling is an unequivocal good. Why would they? Recycling does not explicitly increase taxes, and it does not provoke potential challenges to industry through, for instance, producer responsibility.

And here is the crux of the problem with diversion: by adopting it as a preferred method, municipalities have prioritized meeting local recycling targets, which essentially means increasing the proportion of waste that is designated for recycling. And, as with almost everything, there is the fine print. So, we need to examine more carefully what this actually means. Let's take polystyrene (Styrofoam), for example. Polystyrene is the most difficult material for the KARC to sell, mainly because of its high volume and low weight (Schliesmann 2012). KARC has a contract with an Asset Recovery plant in North Bay, Ontario (approximately 454 kilometres away from Kingston), which receives Kingston's approximately 47 tonnes of polystyrene annually (ibid.). The plant in North Bay super-compresses the polystyrene and ships 60 per cent of it to Korea, and the other 40 per cent to the United States (ibid.). The super-compressed polystyrene is then refashioned into household baseboard products and picture frames, and then shipped back for sale in North America (and presumably then eventually discarded).[7] The fine print reveals recycling to be less a success story than what product producers and the WM industry (both of whom externalize the cost of recycling to consumers) spend millions of dollars trying to convince us of. Further, when we factor in the environmental costs of transportation, it is difficult to argue that the environmental costs of transporting polystyrene across the globe (to be eventually landfilled or openly dumped) are lower than the environmental costs of landfilling the polystyrene locally (or better yet, refusing to allow companies to use polystyrene as liberally as they currently do, refusing to buy products that come with polystyrene packaging, or refusing to buy certain products altogether). On my most recent visit to KARC in October 2019, an employee told me that polystyrene is currently being transported to Indianapolis, Indiana, for reprocessing, which is approximately 626 kilometres further away than the recycling plant in North Bay, Ontario (and approximately 1,080 kilometres away from Kingston).

Furthermore, as the market price for recyclables is constantly fluctuating – and as KARC begins to deal with different recycling brokers that are responsible for transporting recyclables to a processing facility or other recycle traders – the transportation routes of recyclables can vary widely. As of 2012, the closest recycler to Kingston is Waste Logix in Brechin, Ontario (approximately 300 kilometres away), and Waste Logix accepts Kingston's film plastics (i.e., shrink wrap and shopping bags). However, after the waste is shipped to Brechin, Waste Logix then ships the plastics to a Texas company (approximately 2,700 kilometres away) that processes, sorts, and pelletizes the plastics (Schliesmann 2012). Although we lose the trail at this point, it is worth noting that PlasticsEurope (n.d.) estimates that the vast bulk of recycled plastic ends up landfilled. It is this kind of fine print that leads electronics waste scholar Josh Lepawsky to refer to all of this as our "recycling trap" (2018, 159).

## Eco-Efficiency and Waste-as-Resource

Eco-efficiency is a mode of governance in which governmental stakeholders partner with other non-state actors (in this case, WM companies) in order to develop and implement reduction and (more commonly) diversion initiatives. An example of eco-efficiency in Kingston is the organics program. Since 2008, the city has contracted Tomlinson Environmental Services (formerly Scott Environmental Group) to run the municipality's green bin services. The service, under Tomlinson (previously, Norterra Organics), accepts wood, household organics, and pre-industrial and post-industrial organics; they then sell finished compost back to Eastern Ontario (City of Kingston 2015). Tomlinson, by reducing the environmental impacts of organics waste and recovering value from the city's organics materials (compost), illustrates the eco-efficiency mode of governance.[8]

In practice, this industry–government relationship is subject to an interesting set of power dynamics whereby the city needs to contract its WM services to an industry (because waste is privatized in Canada), and industry benefits from lucrative waste disposal contracts with the municipality.[9] With KARC limited in its capacity to house recyclables – and with the entire recycling system dependent on fluctuating economic markets – waste industries are contracted to ship recyclables to buyers across Canada and the United States, and indeed around the globe. While recycling is praised for diverting waste from landfills and incinerators, it does not obviate the need for transportation, as we have seen, because recyclables must still be moved from where they are gathered to a sorting centre that may or may not be located in the same municipality, and then on to one or more materials recycling facilities that may be out of province or in another country (Chertow 2009).

When pressed to discuss where mid- and long-distance waste transportation fit into Kingston's current WM system or future plan, councillors mostly focused on local transportation (i.e., curbside collection) and did not offer, even when prompted, any information about the transportation of Kingston's waste outside the municipality. Although several councillors understood that Kingston's waste was being transported, only three knew where or how far it was being transported (one was the respondent from industry, and the other two were Kingston waste managers). This indicates how the transference of responsibility for waste from contractor to contractor may well undermine city officials and members of the public knowing where, and at what environmental cost, a municipality's waste actually travels. And again, transportation is just one of the limiting factors of diversion.

Finally, waste as a resource mode of governance, whereby EfW and other recovery technologies are prioritized, is largely expressed in the city's search for a waste processing facility that will turn waste into a monetary resource (i.e., electrical or other form of energy). As the previous chapter detailed, since 2006, the City of Kingston has been undergoing an IWM plan to determine the viability of an alternative waste technology. Although the city council has shown interest in a local waste processing facility, citing the opportunity to "turn garbage into a resource," there are still no definitive plans to adopt any technology (see chapter 2). As of 2017, the city remains reliant on "maintaining the status quo of transporting and landfilling" their wastes.

To an even lesser extent, the city also adopts the waste-as-resource mode of governance through periodic Giveaway Days (three in 2017), on which residents are encouraged to put out unwanted items for neighbours to pick up (City of Kingston 2016). This initiative is supplemented through local community-led initiatives, such as clothing swaps or garage sales, which are indicative of the social benefits that the waste-as-resource mode of governance is meant to generate. These non-governmental stakeholders organize such activities outside of any formal governmental infrastructure. Moreover, these activities constitute a very small fraction of Kingston's waste production (recall from chapter 2 that MSW accounts for approximately less than 1 per cent of Canada's waste production – so, the proportion of waste that Kingstonians are transforming into resources through reuse for profit is miniscule indeed). The city advertises Giveaway Days as "just one of the ways the City is working to help meet its goal to divert 60 per cent of household waste from landfill by 2018" (City of Kingston 2016, 1), indicating that although such reuse and waste-as-resource initiatives are an attempt to prevent materials from becoming obsolete, the overall approach is still entrenched within the diversion mode of governance.

Theoretically, there is still much room to grow this form of governance, in terms of green job creation (see Pollans 2017). As Sustainability Kingston, a nonprofit, continues to implement the City of Kingston's 2015 to 2018 Strategic Plan, it seeks to create more environmental jobs for Kingston's citizens (City of Kingston 2016). As it stands, waste-as-resource is hypothetically meant to include social and environmental benefits or resources but is practically used in waste-to-energy technologies that turn a profit for industries. With a continued lack of concrete examples, of the kind of waste-as-resource that Bulkeley, Watson, and Hudson (2007) and Pollans (2017) discuss, in practice this mode becomes economically focused.

## DIVERTING PUBLIC ATTENTION AWAY FROM UPSTREAM ISSUES

The modes-of-governance model provides a valuable framework with which to make sense of the myriad relationships between government, industry, and citizen stakeholders. It highlights the ways in which government (primarily at the municipal level) defines waste as an issue (environmental, human health, political, aesthetic, and economic), and determines how the management of waste is governed. As the analysis in this chapter demonstrates, despite formal and rhetorical declarations endorsing the waste hierarchy's emphasis on reduction, the actual governance of MSW relies primarily on disposal and diversion, with the latter further reduced to recycling. It is much easier to develop and implement institutional policies and practices that do not disturb neoliberal capitalist circuits of production and consumption than to tackle upstream concerns with reducing the quantity (and ever-increasing toxicity – see chapter 7) of waste. And since their remit increasingly covers both waste disposal and recycling, WM industries consistently and forcefully encourage both disposal and recycling because they profit through either mode. They do *not* profit from reduction, reuse, or refurbishment.

Members of the public are asked to participate in their community's WM primarily within the restricted downstream parameters of disposal and diversion, and through the explicit emphasis on the "good environmental citizen." And again, as chapter 2 detailed, this form of governance structures waste as an issue with robust, reliable, and environmentally safe technology, and as the responsibility of the "good environmental citizen." To wit, the city's annual budget for waste-related education increased from $32,000 to $44,000 in 2013 in order to support "continued promotion ... necessary to reinforce good practices and to encourage residents to improve on their practices" (City of Kingston 2013f, 3). As the previous chapter described, the city offers initiatives like Train the Trainer, which provides educational training to student representatives to "help them provide accurate information to tenants when incorrect materials are seen being placed at the curb" (ibid., 37). The city also introduced various educational programs and events, such as the Do What You Can program (a provincial hazardous waste or special waste diversion program) and Pitch-In Kingston (a community cleanup program). The emphasis on habitual (re)education and the surveillance of others serves to remind residents of their responsibility for WM practices: "pleading ignorance is no longer a defense for anyone" because "by now, residents know what is a recyclable and what is garbage" (*Kingston Whig-Standard* 2009, 1).

Unsurprisingly, individual Kingstonians do largely conform their behaviour to the goals of diversion: residents accede to this top-down form of neoliberal governance (Lougheed, Hird, and Rowe 2016). But this has produced new problems. The quantity of recyclables has exceeded the expected quantity, established by the standardized size of the blue box, resulting in the overfilling of recycling boxes (Kingston City Council 2013b). As a result of this overfilling, both wind and sanitation workers strew the streets with the contents of the blue boxes, increasing the amount of litter (ibid.). To deal with this litter issue, the city has (again) adopted a downstream individual responsibility approach, providing a second blue box to residents who request them.[10]

According to the latest report of Kingston's IWM plan (6 March 2019), the "options" that are "best suited for integration within the existing system" are to:

1 increase the cost of garbage bag tags;
2 explore full user-pay options, including a policy or program consideration for low-income residents;
3 prohibit the disposal of recyclables and organics in the garbage stream and enforce the use of clear bags for garbage;
4 reduce the frequency of garbage collection from weekly to bi-weekly for most eligible categories (this option may allow for bi-weekly collection for yard waste);
5 limit the number of additional "tagged" garbage bags;
6 reduce the number of "two bags of garbage" weeks;
7 increase the size of the blue and grey boxes;
8 make the use of green bins at multi-residential properties mandatory;
9 eliminate fees and charges for schools to participate in the green bin organics program; and
10 provide two size options for green bins: an 80 litre or a 45 litre.

Kingstonians should be forgiven for thinking they bear all of the blame (and thus the responsibility) for the city's waste problem. The city's WM plan almost exclusively focuses on individual and household (residential) responsibility for waste. The ten options under consideration all aim to reduce waste through increased recycling and affix all responsibility to individuals and households.

The relationship between the provincial and municipal levels of government plays a central role in institutionalizing disposal and diversion. The provincial legislature is responsible for outlining the terms of a municipality's waste services, and the municipality is charged with ensuring that these terms are met. Municipalities have the option to outsource their WM in its entirety or in part to industry; indeed, outsourcing has become municipalities' modus operandi, further solidifying their dependency on industry (Campanella 2015). According to a city councillor, municipalities want to do more in terms of innovative waste services that are not focused solely on disposal and diversion, but are constrained by the province. In 2010 alone, Kingston received $827,224 from Stewardship Ontario, an organization *funded by the recycling industry*, for meeting its waste diversion targets, which mainly consists of, unsurprisingly, recycling. Put bluntly, because the province focuses on diversion, so too, city council maintains, must the municipalities. Indeed, the *only* financial incentive is limited to recycling. There are signs that this may be shifting, and I will revisit the relationship between provincial and municipal levels of government in the epilogue.

Studies concerned with the material details of recycling clearly show that recycling is not a win-win solution: it isn't even a break-even response. In this chapter, I have focused on waste transportation as a key yet often-overlooked example of recycling's mitigated efficacy. The devil, it may be said, is in the details. Because Kingston does not currently have the infrastructure necessary to divert waste from landfill within its borders, 100 per cent of Kingston's recyclables are shipped outside of the city. According to a greenhouse gas calculator, based on Statistics Canada trucking data, a heavy truck emits 114 grams of carbon dioxide equivalents per tonne-kilometre ($114g\ CO_2e$/t-km) (Government of Canada 2016). A typical waste collection truck will "have a curb weight of 13 tonnes and a payload capacity[11] of 9.5 tonnes, but gross vehicle weights vary up to 27 tonnes" (OWMA 2016, 10). These curbside pickup trucks are the most energy intensive vehicles on the road, due in part to their weight, low speed, stop-and-go cycles, and high idling times (ibid.). According to Environment Canada, MSW collection vehicles are classified as heavy-duty diesel vehicles (HDDV), which is the same classification

Table 3.1 | Emissions from heavy-duty diesel vehicles (HDDVs)

| Activity | Type | Emission | Rate | SD | Source |
|---|---|---|---|---|---|
| Idling | Greenhouse gases (GHGs) | $N_2O$ | 0.08 g/L | | Environment Canada |
| | | $CH_4$ | 0.12 g/L | 72 g/L | Environment Canada |
| | | | 0.13 g/L | | |
| | | | 0.15 g/L | | |
| | Criteria air pollutants (CAPs) | $CO_2$ | 2730 g/L | | Environment Canada |
| | | NOx | 144 g/L | | Lim |
| | | CO | 94.6 g/L | | EPA; Stodolsky et al. |
| | | PM10 | 2.57 g/L | | EPA; Stodolsky et al. |
| Driving | GHGs | $N_2O$ | 0.08 g/L | | Environment Canada |
| | | $CH_4$ | 0.12 g/L | | Environment Canada |
| | | | 0.13 g/L | | |
| | | | 0.15 g/L | | |
| | | | 2730 g/L | | Environment Canada |
| | | NOx | 6.68 g/km | | Lindhjem and Jackson |
| | | CO | 26.6 g/km | | Lindhjem and Jackson |
| | | PM10 | 0.17 g/km | | Lindhjem and Jackson |

assigned to long-haul tractors (Agar, Baetz, and Wilson 2012). Table 3.1 provides estimates of HDDV emission rates for long-haul transportation. As Table 3.1 depicts, the most significant emissions are carbon monoxide (CO) and carbon dioxide ($CO_2$). The rates are also given for nitrous oxide ($N_2O$), methane ($CH_4$), nitrogen oxides ($NO_x$), and particulate matter (PM10) greater than 10 micrograms (µg) in diameter. These greenhouse gas emissions (GHGs) are significant contributors to climate change (Government of Canada 2016). In other words, diversion in the form of recycling that relies on mid- and long-haul transportation contributes to the problem of global warming.

Some of the distances covered in this transportation can be enormous, as we have seen in the case of Styrofoam being shipped to Indianapolis, for which the initial transportation distance amounts to 1,080 kilometres, as well as requiring a separate fleet of vehicles responsible for long-distance transportation. When governing by disposal (i.e., sending waste to a landfill), Kingston's waste is most likely to remain in North America – primarily in Ontario, Michigan, and New York (OWMA 2015). However, when Kingston ships waste as a means of diverting it from landfill, the distance is considerably greater – thousands versus hundreds of kilometres. Before Kingston shipped its polystyrene to Indianapolis, it went to Brechin, Ontario (263 kilometres), then on to Montreal (526 kilometres), and then to South Korea (8,574 kilometres), to be shipped back again to Canadian cities in the form of picture frames sold at dollar stores. Roughly, this amounts to 18,726 kilometres. According to Olmer et al.'s (2017) analysis of container fleet $CO_2$ emissions, container ships alone contribute about 23 per cent of human-produced carbon emissions (more than oil tankers, which account for 13 per cent). In 2015, international container ships produced approximately 31,419 million tons of $CO_2$.

Waste transportation involves several environmental and health-related issues, including climate change due to greenhouse gas emissions, a reliance on non-renewable fossil fuels and a substantial expenditure of energy, and various safety risks incurred through contamination, spills, leaks, and injuries (Thomson 2009; Eisted, Larsen, and Christensen 2009; Gregson, Watkins, and Calestani 2006). The further or more complex the transportation route, the higher the probability that there will be negative impacts on human and environmental health and safety (Thomson 2009; Eisted, Larsen, and

Christensen 2009; Gregson, Watkins, and Calestani 2006). Garbage trucks account for 5.12 human deaths per 100 million miles travelled (Rosenfeld 2013), compared with other light motor vehicles (cars and light trucks), which account for 1.13 human deaths per 100 million miles travelled (Insurance Institute for Highway Safety 2016). Furthermore, waste collectors are routinely exposed to hazardous conditions, including coming into contact with toxic materials, inhaling pathogens and dust, and handling sharp objects (Yong Jeong 2016). Although deaths in the occupation are relatively rare, "ergonomic injuries, such as back strain, are commonplace and cuts from sharp objects and exposure to bacteria and toxins are always a threat" (Tibbetts 2013, 185). According to Tibbetts, "if you're on the job for five years ... I would say it's a pretty safe statement that you're going to get some kind of injury" (ibid.). According to the Canadian Union of Public Employees (CUPE), garbage collection is one of the most hazardous jobs, with injuries afflicting approximately 35 per cent of garbage collectors in Canada each year (ibid.).

Indeed, government policies are only able to champion the reduction of global environmental impacts of landfilling through diversion by obviating the global environmental impacts of transporting (often several times), and reprocessing waste. As such, transportation issues must be highlighted in the environmental costs of recycling (Ali 2002; MacBride 2012; Krausz 2012). The municipality, preoccupied with meeting provincially set diversion targets, contracts private waste industries to provide disposal and diversion services, including waste transportation. Through the RFP process, municipalities are primarily concerned with diverting waste from landfill through recycling, and disposing of the rest of the waste that cannot be diverted. The costs of transporting waste to its final destination are not explicitly included in the city's RFP or the industry's bid. As this chapter has detailed, price is the biggest consideration for the municipality when choosing a WM service industry (based on the points allocated for price). If the city's bidding process for WM services is primarily driven by economics, then transportation will only become an issue if the cost of oil begins to negatively affect the industry's bottom line and consequentially affect the municipality's budget. When oil prices substantially rise in the future, waste industries will need to submit higher bids in order to defray their transportation costs. Otherwise, industries will need to source more local or regional options for disposal.

And although Bulkeley et al. identify eco-efficiency as the result of private-public partnerships, and a waste-as-resource governance as the result of communities reconceptualizing waste, I argue that the opposite is true in the Kingston case study: eco-efficiency, in the sense of reusing materials, is governed by publics while waste-as-resource, in the sense of drastic waste reduction, is governed by industries. If waste-as-resource is meant to prevent materials from becoming obsolete, industry and government are better suited than consumers to make decisions in the manufacturing process that would reduce the consumption of virgin materials. However, this also supposes that private-public relationships do not have unequal distributions of power or vested interests that are focused on economic gains rather than social and environmental gains, as Bulkeley, Watson, and Hudson (2007) contend.

We might certainly argue that eco-efficiency and waste-as-resource modes of governance direct WM towards upstream issues insofar as both modes encourage the continued circulation of materials within society rather than the extraction, and/or creation of new materials. However, with cost-efficiency and profit as the primary underlying motive, both modes are as compromised as the diversion and disposal modes, and in some cases far more so. Although it could be argued that eco-efficiency and waste-as-resource governance is more transformative – insofar as waste is meant to be challenged and reconsidered in new terms and possibilities – it is also more difficult to achieve than disposal and diversion (Pollans 2017). Furthermore, these higher-tier modes require significant buy-in from industry and may require significant monetary investments before changes are seen. They also unquestionably require levels of government that are prepared to forcefully regulate industries that both produce goods, and deal with WM. Even though waste-as-resource frames resources in non-monetary terms, they are often equated with moneymaking schemes (i.e., new technologies that capture energy from waste, but require an incredible amount of waste generation to sustain). Similarly, all four of Bulkeley, Watson, and Hudson's (2007) modes of governance are still entrenched in the neoliberal capitalist framework (in regards to individual responsibility and a focus on economic gains). For this reason, I argue for a fifth mode of governance, one that begins with waste as an upstream issue, and prioritizes critical reflection on overconsumption and an economics based on relentless growth.

# SECTION TWO

## Canada's Arctic Wastes

CHAPTER FOUR

# Canadian Settler Colonial Waste

INTRODUCTION

On a trip to Iqaluit, Nunavut, Alexander Zahara and I stood at the end of Akilliq Road, past the colossal West 40 dumpsite, at the point stretching into Frobisher Bay. As a graduate student, Alex worked with me for two years on a project exploring waste issues in Iqaluit (see Zahara 2015) between 2012 and 2014. It is from this point that local residents launch their boats in the summer months, and tourists arrive from cruise ships that tour the Arctic waters. As it happened, a Zodiac inflatable docked, disgorging some ten tourists with bright orange life vests. After some time, two of the British tourists struck up a conversation with Alex and me. Having determined that this was these tourists' first time in the Arctic, I asked what their impressions were. Both women, without hesitation, exclaimed with dismay and some amount of outrage that they were "shocked" and "very disappointed" with the garbage visible everywhere. They had come to the Arctic expecting a pristine landscape – untouched nature – and had instead found plenty of garbage floating in the water, litter strewn across the tundra, and numerous visible dumpsites.

I share this anecdote because it illustrates a common experience that non-locals have when visiting the Canadian Arctic. Even though I study waste and had read plenty of reports about the waste issues in the Arctic before travelling north for the first time, I was similarly struck by the volume, mixture, and toxicity of waste in Nunavut. Profound waste issues have developed in Canada's Arctic in a very short period of time. Municipalities are struggling with open waste dumps that contain unknown quantities of unknown materials, and

Figure 4.1 | Anti-littering poster, Pangnirtung, Nunavut.
"If you see a bear choking...
　Step 1. Get behind him and wrap your arms around his waist.
　Step 2. Perform abdominal thrusts until the obstruction is ejected ...
　or DON'T LITTER!"

open sewage lagoons that drain onto tundra and/or into water systems, as well as numerous military and resource extraction industry sites where infrastructure and waste (some of it highly toxic) has been abandoned.

The area north of 60 degrees latitude covers some 3.4 million km², or 40 per cent of Canada's landmass. Less than 1 per cent of Canada's population lives in the Arctic – most of whom are Inuit. The eastern Arctic city of Iqaluit is the capital of the Canadian government's largest land claim agreement with Indigenous peoples of Canada, which led to the jurisdictional territory of Nunavut.[1] Iqaluit is not only the territorial capital, but also the only city in Nunavut, Canada's largest and newest territory. In 2011, the population of Nunavut was just under 32,000 (Statistics Canada 2011b). Inuit make up approximately 85 per cent of the territory's population, though this is significantly lower in Iqaluit (City of Iqaluit 2014a). The median age of Inuit living in Nunavut is twenty-six (Statistics Canada 2017). None of the communities in Nunavut are connected by road, and the land area of this territory is nearly 1.9 million kilometres² (Statistics Canada 2011b).

These next three chapters focus on waste issues in Canada's Arctic, and more specifically Nunavut. The main aim of these chapters is to argue that waste issues in Canada's Arctic cannot be separated from the history of settler colonialism. Most Canadians living in southern Canada have little idea of the scale and toxicity of waste in the Arctic, or that most of it has been created and abandoned within just two generations.[2] The present chapter details the history of settler colonialism as it pertains to waste. Specifically, I will explore the ways in which settlers, including traders, missionaries, and government officials such as the Royal Canadian Mounted Police (RCMP), and later Canadian and US militaries, introduced waste and wasting practices to the Arctic and to Inuit: these settlers, indeed, taught Inuit to waste.[3]

Waste is a particularly provocative material object with which to think about settler colonialism because of the important part this object (or, rather, heterogeneous mix of "stuff," as the introduction to this book details) has played, and continues to play, in "excluding certain groups of people from specific social, political, and physical spaces" (Moore 2012, 787). The longstanding association of waste, dirt, and disease with racialized and colonized peoples as a justification for practices of subjugation certainly offers insights into waste as a cultural signifier (see M. Douglas 2007; Kristeva 1982), but in the context of our contemporary environmental crisis, which some capture within the term "Anthropocene," waste takes on, I argue in the final section of this book, a distinct hue. That is, the materiality of waste, as one of the most pressing contemporary global environmental issues, is acutely experienced by people living in the Arctic. As waste materials move northwards, communities have few resources to effectively mitigate this accumulation, much less refuse it.

### WASTE-LAND

The contemporary Canadian government, as well as national and international corporate interests in Arctic resource exploitation, is historically and ideologically shaped, in significant part, by a European Judeo-Christian understanding of land and the imperative of its use (Saul 2008). In Old and Middle English, "waste" referred generally to the environment, and more specifically to uninhabitable land. In the 1200s, the Anglo-French word for waste meant "desolate regions," and in Old North French it referred to "damage, destruction, wasteland, or moor." By around 1300, the Old English word for "waste" meant "a desert, a wilderness," from the Latin "empty, desolate, waste." Into the 1600s, waste still referred to land that was "unfit for use" (ibid.). Land that appeared so desolate and inhospitable was synonymous with wilderness. Wilderness, in short, was wasted land: a wasteland.[4]

According to Judeo-Christian morality, wastelands are not only places of desolation and hostility, but also carry a certain obligation – these are places that humans may redeem through effort, hard work, and conviction. In his *Second Treatise of Government*, John Locke wrote:

> God, when he gave the world in common to all mankind, commanded man also to labour, and the penury of his condition required it of him. God and his reason commanded him to subdue the earth, i.e., improve it for the benefit of life, and therein lay out something upon it that was his own, his labour. ([1689] 2011, sec. 32, chapter 5. Note that I do not endorse the sexist references to God and "mankind," etc.)

According to this ideology, the way to redeem idle land is to transform it – through "man's" labour – into usable, useful, and cultivated land. There needs, in other words, to be a human mark on the land. Land not marked by human toil is land unremarkable, unused, and unusable; again, a waste. It is, indeed, *uncivilized land*. Waste, as John Scanlan reminds us, is about indeterminateness: "the references to places or things that belong to neither one person nor another, its [waste] being *the original condition of nature's chaos*" (2005, 25, emphasis added). Wasted and unused land is *ipso facto* without sovereignty. Scanlan goes on to point out that, from this European Judeo-Christian perspective, a wasteland is effectively witness to human failure to dominate nature. In this sense, Scanlan argues, waste and its opposite – utility and value – is "a way of knowing the material world" (2005, 132). Waste lands are then, by definition, places beyond European Judeo-Christian comprehension. They are places that white settlers are divinely directed to commandeer, cultivate, and control.

Thus, and notwithstanding prospectors' and settlers' dependence on Inuit for survival, navigation, hunting, and labour (Brody 2000; Qikiqtani Inuit Association [QIA] 2010; Paine 1977; Wachowich 1999), many historical accounts describe Inuit as merely "surviving" in the Arctic before settler colonization; according to these accounts, Inuit did not thrive, nor live with the environment, but somehow, against all odds, managed to barely sustain themselves in such an inhospitable wasteland. In other words, Inuit were not characterized as working the land in a useful Judeo-Christian fashion – planting, growing, and harvesting crops and domesticating animals. Indeed, Inuit were sometimes regarded as children in need of instruction and governance. Note, for instance, Fridtjof Nansen's preface to Diamond Jenness's *The People of the Twilight*, written in 1959: "One cannot read this charming narrative ... without getting a deep sympathy for these simple, unsophisticated children of the twilight ... a charming people of happy children, not yet stung by the burden of our culture, not burdened by the intricate problems and the acid dissatisfaction of our society" (in Jenness 1959, v). Although Inuit had indeed already been "stung" by colonization by this time, it is clear that from this perspective, hunter-gatherer ways of living with the land seemed not only primitive, but also idle – making no mark (that settlers recognized), utility, or command of the land rendered it a wasteland and Inuit in need of stewardship and proper governance (Brody 2000).

## TEACHING INUIT TO WASTE

The Arctic's short yet profound settler colonial history involves Euro-Canadians and Euro-Americans teaching Inuit how to waste. Prior to the settlement of *Qallunaat*[5] in Nunavut, the Inuit of Baffin Island (what is known as Nunavut's Qikiqtani Region) were semi-nomadic and relied solely on the land and water for sustenance (QIA 2010). One or two Inuit families together hunted a variety of seasonal animals (for example, caribou, seal, ptarmigan, muskox, and polar bear) for food, clothing, tools, and other necessities. In the summer, caribou skin was used to make summer tents, and in the winter, snow and ice were used to make *iglu*, and sod was used to make houses called *qammaq* (ibid.).

Generally speaking, the history of settler colonialism in Canada's Arctic resembles that of settler colonialism in Canada's South – though, significantly, it occurred over a century later.[6] Early explorers depended on Inuit to survive what Europeans experienced as the harsh northern climate. When the Hudson's Bay Company and other outfitters arrived in the Arctic, they organized hunting around capital and profit rather than subsistence. Inuit visited the area of Iqaluit to fish, hunt, and trade, but retained (relatively) their independence.[7] Even as late as the early 1950s, only fifty Inuit lived in Iqaluit permanently (QIA 2010). However, when the American military selected the region as a Second World War airbase, Inuit began moving into year-round settlements. By the late 1950s, when the Canadian government constructed the Distant Early Warning (DEW) Line as a strategic defence against Soviet invasion during the Cold War (see chapter 5), Inuit began to settle in earnest (Eno 2003; Gagnon and Iqaluit Elders 2002). The shift to a sedentary lifestyle, which was both driven by, and reliant on, government subsidies and a settler-style, labour-based economy, resulted in deep social and cultural changes that transformed the relationship between Inuit, land, family, and community practices, food, health, education, and waste.

Like First Nations and Métis in southern parts of Canada, Inuit were rapidly and purposefully assimilated into mainstream Canadian culture through violent means. Anthropological reports from the mid-twentieth century describe Iqaluit (then Frobisher Bay[8]) as a town where "sophisticated southern populations," "rugged old-timer Northern whites," and a "shadowy social world of [M]étis and [N]atives" (Fried 1963, 59–66) lived together in close proximity.[9]

Whereas most government workers lived in "Southern Canadian type 'suburbias'" (ibid.), many Inuit lived in self-made, one- or two-room shacks (Honigmann and Honigmann 1965). Since the Canadian military did not provide housing for Inuit casual labourers, Inuit used the military's discarded materials for shelter. And while an unnamed American military official described the "ingenuity and cleverness" of his Inuk labourer in preventing waste materials from being produced, this admiration stands out because it was not shared by his colonial colleagues (Gagnon and Iqaluit Elders 2002, 28).

As the introduction and chapter 2 of this book examine, the line between what is considered resourceful and what is considered dirty was (and is) largely dictated by normative assumptions about cleanliness and waste. As Marie Lathers notes, "management of the abject" (i.e., of feces, dirt, or waste) was central to North American settler colonialism (2006, 419). Here, as Warwick Anderson puts it, waste was used to delineate "the polar opposites of white and brown, retentive and promiscuous, imperforate and open, pure and polluting, civilized and infantile" (2010, 170–1). Though many Inuit recall scavenging for food and other materials regularly abandoned in open dumps by the American military (Gagnon and Iqaluit Elders 2002), the Canadian government proved less inclined to allow Inuit to scavenge. Indeed, one government official noted with disgust that when Inuit began wearing Euro-Canadian cotton materials, that the clothing was worn "until ... it fairly rots off" (Lackenbauer and Shackleton 2012, 8).

All government employees were enjoined to "assist" Inuit in their transition to modernity by "insist[ing] upon the maintenance of ... cleanliness and sanitation amongst the Eskimo employees and their families" (ibid., 10).[10] By 1960, for example, scavenging for household materials was banned in Resolute Bay. In Iqaluit, at least one Inuk reported being fearful of getting caught by military officials for scavenging for wood and mattresses at an abandoned dumpsite (Gagnon and Iqaluit Elders 2002). Government reports from the mid-twentieth century discussed the difficulty Inuit had in adapting to settler Canadian standards of waste management (WM) (e.g., Harrison 1964; C. Thompson 1969). Ironically, Inuit were hired to take care of laundry operations, sewage disposal, waste collection, and cleaning for the Euro-Canadian residents, suggesting that while considered unclean themselves, Inuit were entrusted to unburden white colonial settlers from the toil of maintaining their own colonial cleanliness (Gagnon and Iqaluit Elders 2002; Harrison 1964; Lackenbauer and Shackleton 2012).

The forced shift of Inuit ways of life towards (mainly casual and low-paying) wage labour and a market-based economy was reinforced by several directed government initiatives. Examining the development of Iqaluit, Matthew Farish and P. Whitney Lackenbauer note that by the mid-1950s, Euro-Canadian bureaucrats took a "high modernist" approach to development in the Arctic (2009, 520). The explicit goal was to build "a nation in the northern half of this continent truly patterned on *our* [southern] way of life"[11] (ibid., 518; original emphasis). More than this, Canadian government and industrial interests were designing an Arctic that would be a "*safe* space for development projects" (ibid., 523; original emphasis). As more and more Inuit were assimilated into the neoliberal capitalist market economy, Inuit across Nunavut began to rent government-subsidized housing. These houses were unfit for the climate, and most relied on electricity for heating, which was turned off if tenants did not make their rent payments (QIA 2010; Gagnon and Iqaluit Elders 2002; C. Thompson 1969).

The Canadian government's particular mode of racialized paternal governance was (and still is) reinforced in myriad ways. Government officials, military, and southern industry personnel justified the assimilation of Inuit on the grounds that it was necessary not only for northern development initiatives, but for Inuit survival. Put in a non-colonial register, Inuit subjugation was so successful that the traditional Inuit way of life, as well as the people themselves, were in danger of extinction. In 1964, the Department of Northern Affairs and Natural Resources[12] created *Qaujivaallirutissat*, a guidebook written in both English and Inuktitut, which was designed to help Inuit "when they are faced with the many new things which are happening in the North" (Harrison 1964, 8). Here, the Canadian federal government explicitly recognized the increasing reliance of Inuit on the southern market-based economy, stating that the "[Working Eskimo] can no longer hunt with bow and arrow like in the old days ... some of them would die from hunger if they are not helped by the whitemen [sic]" (1964, 10). The Guidebook advised that educating Inuit children with a settler Canadian curriculum was a vital factor in achieving Inuit assimilation:

> Eskimo [sic] children know a lot about the animals, birds and flowers around them. In school they can learn more about the

habits and usefulness of many things in nature. At school he [*sic*] also learns about mines, machines and factories in which many people work. He [*sic*] will learn how useful these things are to all men [*sic*]. (Ibid., 64)

Hundreds of thousands of Indigenous children throughout Canada – including Inuit – were removed from their families and forced to live in residential schools.[13] Many children and youth were also forcibly removed from their families and placed in white settler families in southern Canadian communities. Many Inuit believed (often accurately) that their family allowances would be taken away if they did not send their children to residential schools – this would have meant a loss of crucial food or housing needed to support small children and elderly relatives (QIA 2010). Like other Indigenous peoples across Canada, Inuit were violently subjected to a plethora of means ostensibly designed to "civilize the savage" – from assigning Inuit numbers instead of names (then assigning white colonial names when the numbering system failed) to beating Inuit children who spoke Inuktitut and thus isolating subsequent generations from their Elders, to literally killing Inuit means of feeding themselves through the systematic slaying of *qimmiit* (sled dogs) (see chapter 6) thus fast-tracking Inuit's incorporation into the capitalist labour market, and many more official measures. As Inuit men and boys were forcibly assimilated into low-paying and casual wage labour, girls and women were being forcibly assimilated into settler colonial roles as housewives and mothers, both roles bearing responsibility for family "morality."

Indeed, educating Inuit youth (particularly young girls) in southern communities was thought to have the added bonus of influencing Inuit women to become better at household chores (C. Thompson 1969). Inuit women, who were traditionally the dominant figure in the tent household, but whose "authority seems to have been usurped" by settlement (ibid., 20), often found the adult education classes organized by the Canadian government to be both boring and demeaning. Similar to American colonialists in the Philippines, which, as W. Anderson (1995) points out, quite literally examined slides of Indigenous people's feces, Canadian federal government employees were sent to inspect the cleanliness of Inuit homes. These government officials noted that Inuit women's housekeeping "lacks organization" (C. Thompson 1969, 13). Federal government officials recorded notes on Inuit diet ("almost all the food was bought from the store"), patterns of food preparation ("[s]oups are heated but do not always have water added to them"), shopping ("men make most of the purchases"), and cleanliness ("toilet bags are allowed to fill before they are removed ... Washing clothes is still a problem in many homes") (ibid. 23, 17, 27, 14–15). So, in the late 1960s, adult education classes were provided to Inuit women whose housekeeping did not "measure up to the standards set by white women" (ibid., 23). One government official, at least, recognized that the decision-making roles of Inuit women in families and communities had been irrevocably changed. His solution was to recommend involving Inuit men in the household, mainly as a way of enforcing settler colonial gender roles: "The influence of the men on purchasing, cooking, and home care should be realized and exploited ... since women have been excluded from some of their traditional decision making situations" (ibid., 30).

In 1999, through the political struggle of Inuit activists (who themselves were survivors of Canada's residential school system), the Nunavut Land Claims Agreement was made into law (for the toll this took on the Inuit negotiators involved, see Arnaquq-Baril's 2009 documentary film *The Experimental Eskimo*). The result was the creation of the largest land claim in Canada's history and Inuit self-governance over the newly formed Nunavut territory.[14] Inuit were no longer the "eaters of raw meat," nor were they the identification number given to them by federal government officials;[15] now Inuit were to be considered "real human beings,"[16] with final decision-making authority over the territory's government and development.

Since its colonization, both the population of Iqaluit and the amount of waste it produces have grown rapidly. In 1989, when Iqaluit's population reached nearly 3,000, the city was producing approximately 15,000 cubic metres of waste annually (Heinke and Wong 1990). Plastics, which comprised only 4.2 per cent of the waste stream in 1974 (compared to 10.1 per cent in the rest of Canada), increased to 13.3 per cent by 1989 (ibid.). In 2011, with a population of just over 7,400, annual waste production was calculated at 82,805 cubic metres (Exp Services 2011, 5). As Inuit activist and writer Zebedee Nungak wryly notes, "Now, our garbage is as 'civilized' as anybody else's" (2004, 18).

## WASTE AND SOVEREIGNTY

In the European tradition, waste and sovereignty are intimately connected. Sovereignty is based on land ownership and land use rights.[17] As Rob Huebert explains, sovereignty consists of "a defined territory; an existing governance system; and a people within the defined territory" (2011, 14). As historians have noted, in the late 1800s and early 1900s, the Canadian government's approach to the Arctic was largely one of "absent-mindedness" (Shackleton 2012, 12). But by 1925, as Janice Cavell and Jeff Noakes argue:

> Canada's haphazard Arctic policy had been transformed into something much more clearly thought out. Canadian officials knew exactly what they wished to claim and how, and they had defined the boundaries within which they intended to work. (2010, 9)

At the time, this shift was more the result of European northern exploration and the Canadian government's desire to officially establish the Arctic as part of Canada than it was about the government's consideration of the land as especially usable or resource-laden. Indeed, descriptions of the Arctic at this time most often called attention to its absolute barren state. For instance, a British Colonial Office statement from 1879 reads: "The object of annexing these unexplored territories to Canada is, I apprehend, to prevent the United States from claiming them, and not from the likelihood of their providing *any value* to Canada" (in Marcus 1992, 57; emphasis added).[18]

Thus, in the early days of European exploration of the Arctic, explorers, traders, government officials, and, eventually, media correspondents described the Arctic as an uncharted, unknown, and inhospitable wasteland. For instance, a *Life* magazine correspondent wrote in 1963 that the Arctic "might as well be space" and that "flesh freezes solid in 30 seconds" (Mydans 1963, 27). Even as sovereignty concerns heightened, and the Canadian government began to think of the Arctic in terms of sovereignty and resources, the territory was largely idealized as a frontier to be confronted and overcome (Lackenbauer and Farish 2007). The RCMP's involvement in the Arctic, and later, that of the military, through the occupation of outposts and eventual construction of the DEW Line, is commonly described in glowing terms, of human ingenuity and strength in taming the barren wilderness. Or, as Lackenbauer and Farish put it, "the construction of radar lines and associated settlements undermined the perception of Canadian wilderness as inhospitable" (2007, 923).

For Inuit who had already been living in the Arctic for millennia, the territory was plentiful in sustenance. A rich and growing literature documents the intimate and highly knowledgeable relationship Inuit have with the territory. The Qikiqtani Inuit Association (QIA), an Inuit-led group working to hold the federal government accountable through an inquiry into justice issues stemming from the mid-twentieth century (led by the QIA) proclaims: "Throughout the Arctic, all Inuit experienced a very long period in which the sea and the land provided almost everything people needed" (QIA 2013b, 14). Inuit developed a complex and intimate familiarity with their regional hunting grounds:

> Every geographic feature ... has names and the name is a metaphor for the totality of the group remembrance of all forms of land relatedness, of the successes and failures in hunting, it recalls births, deaths, childhood, marriage, death, adventure. It recalls the narrations and the ancient sanctified myths ... As Inuit travel across the land, sea, and ice, they strengthen and deepen their relationships with each other and deepen their understanding of their own pasts and kin. (Williamson, in QIA 2013c, 13)

It is within this ideological and material context that the comparatively short, yet profound, colonization of the Canadian Arctic took place. In what the Qikiqtani Truth Commission calls the "official mind of Canadian colonialism" (QIA 2013f, 14), Inuit were initially, as I have said, the subject of intermittent interest, as primitive but benign people barely surviving in an unimaginably harsh wasteland. In the 1950s, this view transformed as Canadian politicians became more interested in increasing Canada's sovereignty over the Arctic through Arctic claims, and later seeing it as a potential source of resource development and profit. In 1953, then prime minister Louis St Laurent declared that a new Department of Northern Affairs and Natural Resources would officially administer over the Arctic and its peoples (who were already assumed to have been under Canadian authority).[19] Where once Inuit had largely thrived within their own territory, they were now increasingly seen as physical bearers of

Canadian sovereignty, with whole families violently displaced at times to areas that the Canadian government wanted to claim as Canada (Tester and Kulchyski 1994). As such, Inuit were increasingly governed through trade with the Hudson's Bay Company, the RCMP, and other Canadian officials.

The RCMP were pivotal mediators in implementing the increasing number of regulations and policies issued from a rapidly expanding southern bureaucracy concerned with Arctic affairs. As Ryan Shackleton observes, "The evolution of the RCMP in the Eastern Arctic is inextricably linked to the evolution of the North. First seen by Inuit as a mixture of benefactor and punisher, the RCMP were Canada's closest tie to the Inuit of the Eastern Arctic" (2012, 19). The RCMP's main directive was to monitor Inuit, collecting census information and vital statistics on the population. This was a complicated relationship: on the one hand, the RCMP were "responsible at times for relieving starvation, soothing illness, and saving lives (ibid., 9); whilst at the same time the RCMP were also charged by the Canadian government with enforcing the law, investigating and adjudicating complaints, collecting duties and liquor permits from traders, and monitoring and controlling animals, including *qimmiit* (sled dogs) (as chapter 6 will explore). As Shackleton notes, it was "not uncommon for the police to use coercion and intimidation when dealing with Inuit" (2012, 11), and Inuit remember feeling both admiration and apprehension towards the RCMP (QIA 2013b). Some of the RCMP's attitude towards Inuit as child-like and primitive led to an inconsistent focus on self-reliance and (what the RCMP defined as) welfare handouts. The movement of Inuit into Arctic settlements has been variously analysed as an intentional and devastating, sometimes brutal, displacement (Tester and Kulchyski 1994), and as a voluntary migration of Inuit in search of more hospitable environs (including for hunting, and for southern attractions such as housing, welfare, and labour) (Damas 2002; see also Alunik, Kolausok, and Morrison 2003).[20]

### WASTE AND SECURITY

In differentiating sovereignty from security, Huebert writes: "the core issue of Canadian Arctic sovereignty is control; the core issue of Canadian Arctic security is about responding to threats" (2011, 21). In the contemporary Arctic context, security has come to focus on a range of issues, which have also become concerned with responding to environmental threats as well as maintaining Canada's Arctic borders (Griffiths 2011; Lackenbauer 2011b). Canadian Arctic security is closely linked with military involvement, whether in terms of the construction and operation of the DEW Line (see chapter 5), the various military operations launched in the Arctic such as Operation NANOOK and Operation NUNALIVUT, or the potentially emerging Arctic security regime that involves enhancing "Canada's military presence and capabilities in the Arctic" through various measures including the increased presence of Canadian military personnel and infrastructure (Lackenbauer 2011b, 229).

There is, at the very least, an unresolved conundrum here between the military's increasing mobilization around environmental disasters, and its own profound contribution to environmental degradation and devastation (see chapter 5 for an inventory of some of the known contaminated sites in Canada involving the military). Referring to the rapidly expanding military presence just south of our border, Shiloh Krupar notes that "the U.S. defence industry is one of the biggest polluters on the planet and a main contributor to Superfund's National Priority List; it now partners with private industry both to clean up after itself, fueling a multibillion-dollar remediation industry, and to preserve the nature it enclosed with arsenals, airforce bases, and military camps" (2013, 38). As the first section of this book has explored, the privatization of WM means that the Canadian government is heavily dependent upon industry to deal with our increasing amount and diversity (numbers and kinds of contaminants) of waste. The Canadian military's partnership with industry is already well known to involve military operations and their cleanup, as the examination of the DEW Line construction, operation, closure, and remediation in the next chapter will show.

## TRUE NORTH STRONG AND FREE

At the 2009 G20 Summit, former prime minister Stephen Harper claimed that Canada had "no history of colonialism" (Ljunggren 2009). In this contemporary refrain of Canada's settler colonial discourse, Harper reiterated the government's wilful obtuseness that Canada is a nation whose resources and opportunities are shared equally by all citizens. As well as denying hundreds of years of Old World dependence on Canadian resources, and in turn of prospectors' and settlers' dependence on Indigenous peoples for survival, navigation, hunting, and labour – all of which inspired the material and cultural subjugation of Canada's original peoples – Harper's statement exemplifies the liberal state that Foucault identified, which "justifies its jurisdiction on a type of origin myth" (in Asch 2007, 281). Faced with the long and deep history of Indigenous peoples in what became Canada, Michael Asch argues that settlers chose to identify settlement as the historical starting point of sovereignty. The Crown declared Canada *terra nullius* before colonization – an absurdum recently reiterated by the Supreme Court of Canada (for a detailed critique of this term in the context of Canadian and US settler colonialism, see Mackey 2016).[21]

Embedded in the colonial imagination of sovereignty is the messy juxtaposition of the Arctic as simultaneously the "true north strong and free"[22] (a remote and pristine landscape whose innocent history embodies an aesthetic of uncontained and uncontaminated wilderness); the Arctic as Canada's largest and most diverse emerging resource for industrial extraction – a vital piece of the circumpolar pie – and, increasingly, the North as an anthropogenic trace and therefore "a symbolic pinnacle for global sustainable development" (Shadian 2006, 249).[23] The dramatic increase in demand for northern natural resources over the past twenty years has only intensified with the prospect of climate change making these resources more accessible (Southcott 2012). According to Aboriginal Affairs and Northern Development Canada (AANDC), the Arctic contains about 25 per cent of Canada's remaining discovered recoverable crude oil and natural gas, and about 40 per cent of Canada's projected future discoveries (Government of Canada 2010). This means more people and equipment moving temporarily from south to north, and much more drilling and extraction – inevitably leading to more waste.

A technocratic language of environmental management is increasingly eclipsing debates about Inuit control over land and sea; a discourse that includes not only the interests of scientists and conservationists in land stewardship, climate regulation, and biodiversity, but those of the oil and gas industry, mineral mining industry, tourists, and other settler Canadian and international corporates. As such, Inuit rights are being fused with resources into a single issue (Shadian 2006, 250). This means that Inuit rights over Arctic development in areas of oil and gas exploration, hunting, and fishing is now advanced on the grounds of thousands of years of successful Inuit stewardship. However, this stewardship is formulated in terms that assume resource development as a given – and moreover, as Jessica Shadian argues, within a discourse that corroborates Canada's ongoing settler ideology (ibid.). In other words, Inuit have rights because of their status as Canadians, and it is the needs of Canadians as a whole (which, according to neoliberal capitalism, consist of resource extraction, profit, and global corporate investment) that define the terms of sustainable development in the Arctic. As the Canadian government's "Northern Strategy" states:

> Canada's North is a fundamental part of Canada – it is part of our heritage, our future and our identity as a country. The Government has a vision for a new North and is taking action to ensure that vision comes to life – *for the benefit of all Canadians*. (Government of Canada 2014, emphasis added)

This means that "Indigenous people have an [*sic*] effect been engaged in a massive program of foreign aid to the urban populations of the industrialized North" for the past several hundred years (Kloppenburg 1991). Through contemporary interests in development, Inuit communities are being assimilated into greatly expanding industrial corporate interests in the Arctic through resource extraction, casual labour, capacity-building training, and the tourist trade (Bravo 2006). Inuit communities are themselves caught up in often fraught internal struggles as they negotiate access to resource development (e.g., Bernauer 2012; CBC News 2015).

For Inuit, the Arctic is home: a place of complicated histories of violent subjugation, collective memory, landscape, survival, intergenerational tradition, and more. Researchers continue to identify the multitudinous human health and environmental risks that attend northern development, such as living with contamination (see for example Kafarowski 2004; and Sandlos and Keeling 2012). Those with strategic interests in the Arctic, first as a military defence site in the 1940s and then as a site for resource extraction that has been accelerating since the oil crisis in the early 1970s, continue to promise the spoils of Western civilization to Inuit: more jobs, more training, more money, and greater investment. However, as one advocate for Inuit rights challenging the Canadian polar gas pipeline project points out:

Initiatives such as the pipeline have too often been proposed together with promises that it will shepherd native people into the 20 [sic] century ... [instead] too often it serves only to dislocate and disorient native peoples and leaves them unequipped for the 20th century, stripped of their lands and waters and the ability to follow their traditional pursuits once it has passed them by. (J. Bayly, in Shadian 2006, 253)

Thus, amongst whatever dividends Inuit may or may not actually accrue – and studies demonstrate that many are peripheral and temporary, and come with problems associated with what is known as the "staples trap" or "resource curse" (see Southcott 2012, for example) – northern development ultimately leaves substantial waste in its wake:

Today, the greatest and certainly the most direct threat to the security of Arctic residents stems from damage to the environment. The Arctic, in effect, has been treated as a dumping ground by governments, military establishments and industries concerned only with the needs of southern societies. (Simon 1988, in Shadian 2006, 256)

### SELF-DETERMINATION

This is not to say that Inuit do not act, know, or care about WM. As chapter 9 will detail, in Iqaluit, as with other Arctic communities, there is a multiplicity of perspectives on waste and other issues. Rather than rehearsing a familiar settler colonial discourse that defines (and thus confines) Indigeneity to local or traditional practices and epistemologies, Alex and I argue that Inuit perspectives are embedded in (whether deeply aware of, occurring in response to, or independent of) Nunavut's recent and ongoing settler colonial history.[24]

Many Inuit, for example, consider community and resource development as a necessary and even desirable way forward in the context of Nunavut's myriad social issues – which stem from decades of settler colonial violence (Tester and Irniq 2007; Cameron 2012; QIA 2013a) and "chronic underfunding" from Canada's federal government (Ehaloak, in LeTourneau 2014, 1). Inuit living in Nunavut have among the lowest household incomes in the country (Statistics Canada 2012b), and the difference in average income between Inuit and *Qallunaat* is striking: According to the Basic Income Canada Network, the average income for Inuit in Nunavut in 2014 was $19,000; for non-Inuit it was $86,600.[25] Nearly 60 per cent of those living in Nunavut smoke, ranking Nunavut the highest territory or province per capita in the country (Statistics Canada 2013a). Residents also have high levels of diabetes, heart disease, and other diet-related illnesses, and Nunavut households experience food insecurity at a rate seven times higher than the Canadian average (Department of Health and Social Services 2007). In the grocery store, food comes highly packaged and/or nearing its expiration date. Violence against women is a major issue, and the single women's shelter struggles to meet the overwhelming need. Substance abuse is a major problem, and suicide rates are forty times higher in Nunavut than in the rest of Canada (Chachamovich et al. 2013).

In the discussions that Alex and I had with Iqaluit residents about waste issues in the Arctic, current and historical relationships with the federal government were frequently brought to the fore. For example, Iqalummiut complained of the yearly military exercise Operation NANOOK, in which millions of dollars are spent "defending Arctic sovereignty" rather than addressing "a real emergency," such as the Iqaluit dump fire (CBC News 2014e). Operation NANOOK in November of 2018 involved:

> 2000 Canadian Army Infantry, 1000 Canadian Rangers, and 200 members of the Canadian Coast Guard in the vicinity of Resolute Bay, Nunavut. In addition to these personnel, the Halifax-class frigate Vancouver, the Kingston-class coastal defence vessels Chicoutimi and Glace Bay, 3 x CC-138 Twin Otter aircraft, 3 x CH-146 Griffon helicopters, 2 x CH-147 Chinook helicopters, 1 x CC-130 Hercules, and 1 x CC-177 Globemaster cargo aircraft will participate. If needed, Canadian icebreakers will remain on standby.[26]

To wit, Alex calls attention to the stark fact that the Canadian government spends millions of dollars on a military presence in Iqaluit but refused to contribute money to putting out the 2014 dump fire (Zahara 2018). The Canadian government's interest in the Arctic, as demonstrated through this investment strategy, is focused on keeping Arctic industrial interests in the hands of the Canadian government, while simultaneously preparing for the environmental disasters that this industrial extraction and development have already, and will in the future, precipitate.

Inuit lay bare the relationship between waste and the federal government's "high modernist" project in the Arctic. As one Inuk man commented:

> It's a catch-22 kind of thing ... Because we didn't need television, we didn't need rifles, we didn't need snowmobiles. We were living just fine the way we were. And this white man comes "oh you need shelter, oh you need furniture to get status in your life. Oh you need pots and pans." But we didn't. We were fine the way we were ... As soon as the white man said, "you need to be in communities" ... We were all scattered all over the place, and then the government said we got used to money. And the government said "if you want more money you gotta send your kids to school [in the South]." And that's how the communities formed.²⁷

This resident's characterizations of federal government relationships, as they relate to personal experiences of settler colonialism, are important. They indicate how Inuit and other Iqalummiut participate politically as activists, politicians, industrial negotiators, disengaged citizens, and so on. As another Inuk resident remarked:

> Even on the land – you know, we put them [garbage] in the boats, our tents, [when] we go on the land. Garbage. Garbage. We eat, eat, eat. Garbage. Put it in the garbage bag. Right after, if we are going to leave from the campsite, we are going to take the garbages too. Because you have to respect the land. It's just the land and the nature. I know nature gets mad all the time. We cannot handle it. Right? Human beings. Humans, us humans, cannot [pauses] like [pauses] control the nature. Right? ... We live in the richest land of Canada and of the universe. We have sapphires, we have crystals, and animals ... There's a lot of beautiful lakes, you can catch some fish. Red fish, Arctic char, you know? Other stuff. Salmons. We've got all of them almost. Arctic cods. You can go like five minutes and go find cod, and go make some fish and chips for ourselves ... That's the way. We don't deal with garbage. We [Inuit] don't want them [the garbage] to be in the lakes, or on the river, or on the sea, because we have to eat it [and] be

responsible for everything ... I know they [the federal government] come and say, "yeah, I'm just a number" and blah blah blah. They don't consider us as human beings. I know it ... They [the federal government] think they found us, but no. Been there for centuries and stuff ... They [my ancestors] lived environmentally free. It was strong stuff. Healthy. You never seen garbage. Nothing.[28]

This statement was made in the context of Inuit struggles for self-determination: it points to a settler colonial cosmology that separates humans from their environment as the cause of anthropogenic "mega-problems," including climate change and other waste-related issues. This resident's use of the term "environmentally free" problematizes the very immaterial construction of "environment" as a concept – of a nature that is placed "outside of" human existence. And the focus on garbage as not having existed prior to colonial contact underscores this point because any materials that were wasted could not have been "out of place," in that the very definition of materials as "out of place" is derivative of a settler cosmology.[29] Moreover, his assertion that we cannot "control" or "handle" nature challenges settler colonial understandings of sovereignty wherein a mastery over people and nature are implicit. Read this way, this is less an aesthetic statement than it is naming the very relationship that the discourse about the Anthropocene has just now discovered: that human/nature relations are inextricable, a theme I will examine closely in section 3 of this book.

Inuit leaders are faced with myriad challenges in attempting to negotiate with the Canadian government, as well as with national and international corporations, for employment opportunities, capacity-building training, much needed infrastructure investment, community investment, tourist trade, investments in health, and environmental protection (Bravo 2006). Inuit leaders such as Nellie Cournoyea have been pivotal in securing Inuit companies' participation and leadership in natural resource development initiatives (CBC News 2016).

Given that corporate interest in the Arctic is almost certainly going to increase, Arctic scholars, Inuit representatives, the media, and government alike tend to argue for both the further formation of international and national regulations regarding development, and their careful monitoring. Lackenbauer suggests that, "overall, the government's commitments to invest in more military capabilities for the North are reasonable and proportionate to probably short- and medium-term threats" (2011a, 106). These threats include climate change, which is precipitating melting permafrost and sea ice, as well as various pollutants, including contaminants of emerging concern that migrate northwards to be consumed by flora and fauna, and, ultimately, local people (Danon-Schaffer 2015). And, of course, "increased military capabilities" in the North means more contamination and more waste: we are creating the very threat that we are attempting to protect ourselves from.

Inuit leaders are highly active in their attempts to protect their communities and the Arctic environment. In two recent declarations, "A Circumpolar Inuit Declaration on Sovereignty in the Arctic" (ICC 2009a) and "A Circumpolar Inuit Declaration on Resource Development Principles in Inuit Nunaat" (ICC 2009b), Inuit from Arctic countries are unambiguous about their sovereignty, and the principles upon which they want northern resource exploitation to take place:

> Inuit desire resource development at a rate sufficient to provide durable and diversified economic growth, but constrained enough to forestall environmental degradation and an overwhelming influx of outside labour.

Many of these development principles explicitly refer to waste issues, including the reclamation and recovery of habitat and affected lands and waters in a way that is thoroughly planned and fully funded; a complete environmental assessment after a project has been completed or abandoned; zero-volume discharge onto land and into Arctic waters; the prevention of spills offshore and eliminating the release of toxic substances; the establishment of an international liability and compensation regime for contaminants of lands, waters, and marine areas resulting from offshore oil exploration and exploitation; and the adoption of the precautionary principle and the polluter pays principle.[30]

## CONCLUSIONS

In their formative article on the effects the Cold War had on militarizing the Canadian Arctic, Lackenbauer and Farish (2007) make the point that the military, through Cold War operations such as the

DEW Line, opened up the Canadian Arctic to emerging state sovereignty and security objectives: military involvement was key to the modernization of the Arctic itself. The authors detail how the military was instrumental in shifting perceptions of the Arctic, noting, "the militarization of northern nature has been flexible enough to accommodate varying discourses of defense, protection, and security" (ibid., 942). In constructing the DEW Line, for instance, the military effectively "undermined the perception of Canadian wilderness as inhospitable" for southern Canadians (ibid., 923). And although the perception of the Arctic as a "frontier land" has not entirely diminished, Lackenbauer and Farish argue that the heavy military funding of, and involvement in, postwar scientific expeditions eventuated in the development of a "delicate Arctic" (ibid., 933) narrative from the Pierre Trudeau years onwards, leading to initiatives such as the 1990 Green Plan and the Arctic Environmental Strategy (1991–96).

It is the complex juxtaposition of oppositional and protective discourses (which include, for the authors, the fact that military operations have increasingly taken place alongside scientific studies in the Arctic) that lead Lackenbauer and Farish to stop short of characterizing the military's rendering of the Arctic as a "sacrifice zone," similar to ways in which the Nevada and New Mexico nuclear test sites have been defined (Krupar 2013; Masco 2006). Whether or not military activity in the Arctic has devastated the peoples and landscape to a degree and kind sufficient to warrant this classification, Lackenbauer and Farish are unequivocal in acknowledging the devastating effects of the military: "Cold War military activities in the Canadian North ultimately constitute part of a global 'treadmill of destruction' tying militarism to environmental and political injustice"[31] (2007, 942). Referring to the Arctic, they observe:

> These military mega-projects radically transformed the human and physical geography of the North. Bulldozers tore permafrost off the ground disrupting ecosystems and creating impassable quagmires. Forest fires, logging, over-hunting, and over-fishing depleted resources in the region. Arriving workers brought diseases, from measles to VD [venereal disease], which devastated [I]ndigenous populations. (2007, 925; see also Kafarowski 2004; and Sandlos and Keeling 2012)

The dramatic increase in demand for northern natural resources over the past twenty years has only intensified with the prospect of climate change making these resources more accessible, by way of, for instance, opening up the Northwest Passage to year-round shipping, and the fact that these resources are running out in other regions (Southcott 2012). Foreign companies such as British Petroleum (BP) are making enormous bids for exploration rights, and countries such as China and India are seeking non-Arctic observer status at the Arctic Council (Griffiths, Huebert, and Lackenbauer 2011). Much depends on a number of factors that remain uncertain at this time: what proportion of what resources (oil, gas, gas hydrates, minerals, and so on) will be uncovered, at what costs, with what developed technology, and in what economic market climate?

These unavoidable uncertainties also mean uncertainties in the amounts and kinds of waste produced as a result of this industrial and military push in the Arctic. In simple terms, when more people and equipment move temporarily from south to north, much more drilling and extraction occurs – inevitably resulting in more waste.

Industry, the Canadian government, and some scholars suggest that resource development may provide much-needed employment for young Inuit (Lackenbauer 2011b). However, Huebert cautions:

> But at the same time, these opportunities may result in serious problems. It is already known that megaprojects often result in serious social problems such as drug and alcohol abuse. Increasing suicide rates are also often associated with societies in transition. Improvements in the economic security of the region may come at the cost of societal security. (2011, 22)

Thus, as Inuit are well aware, the ubiquitous waste that litters the Arctic landscape is the by-product of a settler colonial past, and the prospect of far more waste generated through northern resource development characterizes what Gregory (2006) refers to as Canada's "colonial present," and its forecasted future. Military, mining, oil and gas extraction, and myriad other forms of waste constitute an anthropogenic legacy and capitalism's profound fallout, underwritten by "ideologies and discourses that facilitate resource development and environmental transformations" (Keeling and Sandlos 2009, 123). In significant ways, contemporary WM (or its lack) is a manifestation of Canada's settler colonial tradition, and its implementation requires an assimilation to predetermined neoliberal market-based definitions of what waste is and how it should be managed. As the first section of this book detailed, for most southern Canadian communities, waste is out of sight and out of mind. In this way, waste, both conceptually and materially, marks the success of the settler colonial project – its proliferation and techno-management are predicated upon a settler cosmology that emphasizes dominance over nature. In the Canadian Arctic, this equates to teaching Inuit, in the first instance, to produce waste in new magnitude and kind, and then to adopt neoliberal ways of dealing with waste's proliferation – that is, waste as profit.

CHAPTER FIVE

# Arctic Wasteland

## INTRODUCTION

As the previous chapter detailed, Canada's Arctic is a site of shifting representations: it represents a largely neglected wasteland, commitments to sovereignty and international security needs, contemporary tangled discourses of resource exploitation and Canadian energy interests, Inuit self-determination, and Anthropocene markers of a vulnerable planet in need of human stewardship. These discourses are woven from the fabric of a short but intense and devastating history of colonization, nation building, and a return to Indigenous self-governance within the context of intense and ongoing settler colonialism. The previous chapter focused on providing the context for Nunavut's contemporary waste issues by focusing on the history of wasting practices introduced by settler colonialism as a by-product of the crusade to both "civilize" Inuit and secure the Arctic for industry and the Canadian government's gain. The current chapter details the waste fallout of this historical settler colonialism, as well as from contemporary resource extraction, and military installations and interests. This chapter begins by detailing Iqaluit's waste sites – as Nunavut's only city – before moving geographically outward to further waste sites in Nunavut. It then focuses on the construction and incomplete remediation of the Distant Early Warning (DEW) Line, as one of the very few Canadian military sites for which there is publicly available data.

Part of this chapter is based on research that I conducted with Alex Zahara (2015). With generous funding from a grant I earned from the Social Sciences and Humanities Research Council of Canada, for three months in 2014, Alex lived in Iqaluit, talking with Inuit and

*Qallunaat*, reading archival documents, participating in community events, taking photographs of dumpsites, and so on. Together, Alex and I examined waste within the wider context of settler colonialism, as well as contemporary neoliberal governance practices, to argue that waste is part of the colonial context within which Inuit continue to live: waste, in other words, has become a particular symptom of settler colonialism (W. Anderson 2010). Settler colonial governance leads to the configuration of waste as a fallout of neoliberal capitalism – an unanticipated supplement – which can be managed as a technological issue (bigger and better waste facilities) and individual responsibility for diversion (primarily recycling) (see chapters 2 and 3). More than this, the "failure" of waste in Canada's Arctic communities to conform to settler colonial governance lays bare the ongoing colonial relationship between Inuit and the Canadian government.

## NUNAVUT'S WASTE[1]

In 2014, a four-storey pile of waste, known locally in Iqaluit as the West 40,[2] spontaneously caught fire (Varga 2014a). Although I will analyze this particular event thoroughly in chapter 9, it is worth providing some detail here, in this survey of Iqaluit's waste, because it was of considerable concern to local residents. The dump is reported to have self-ignited on 20 May, the result of bacterial metabolism of the dump's abundant organic and inorganic material. The Iqaluit dump had spontaneously ignited several times prior to the May 2014 fire. Fires in September 2010 (which lasted for thirty-six days), January and December 2013, and January and March 2014 are thought to have contributed to the May 2014 fire. The colossal dump rests atop a peninsula that extends well into Nunavut's Frobisher Bay. For over three months the fire burned continuously. As chapter 9 will detail (and see also Zahara 2018), local residents filed numerous formal and informal complaints to the city regarding the smell of dump smoke. As well, a territorial health department advisory warned that children, women of childbearing years, pregnant women, the elderly, and those with respiratory issues should avoid breathing in dump smoke entirely (presumably the healthy post-menopausal, pre-elderly woman was safe) (Department of Health 2014a).[3] The local elementary school shut down twice due to children complaining of headaches, and several major community events were postponed, including, ironically, the city's annual spring cleanup.

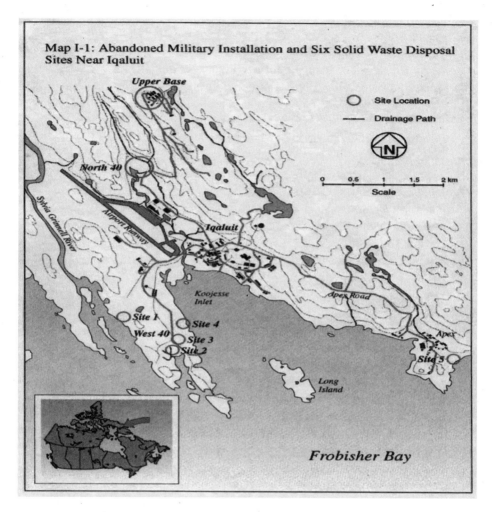

Figure 5.1 | Abandoned waste sites in Iqaluit, Nunavut.

All of this culminated in the federal and territorial governments' reassurances that the public's exposure to dump smoke was safe for human health. Or, at least it should have been safe if standing at a distance of seventy metres from the dump. Meanwhile, a hired landfill consultant explained to Iqaluit City Council that their WM operations "virtually guaranteed this problem would happen" (Sperling 2014, 57). Without technical intervention, the consultant warned city council, the fire would burn for at least another year. The Canadian government refused to financially contribute to extinguishing the dump fire, citing Nunavut's sovereignty. For its part, the Nunavut government declined to financially assist the city due to lack of funds. Putting out the dump fire cost the city nearly $2.75 million, and took over two weeks to complete.[4] During the process of extinguishing the fire, a steady flow of landfill runoff (a combination of water and leachate) was produced, and – purposefully or otherwise – discharged into Frobisher Bay – a process summarized by locals as "dilution is the solution to pollution."[5] Unlike the modern sanitary landfills of Canada's other capital cities, Iqaluit's municipal solid waste (MSW) site is (noticeably) in constant exchange with air, land, and sea: what in settler colonial terminology is typically referred to as "the environment."

The dump is one of the many waste sites located near or within the City of Iqaluit. In January 2013, a 1995 map of the city's contaminated waste sites resurfaced in the local Nunavut newspaper (see figure 5.1). In the newspaper article, then city councillor Terry Dobbin asked federal and territorial politicians to help clean up the community's six remaining waste sites. Although most (if not all) of these sites are left over from federal government military and resource development initiatives, the federal government, military, and industry's consistent silence with regard to their responsibility in cleaning up their own waste means that the immense cleanup is left to the people of Iqaluit, and Nunavut more generally.

To wit, in the over two decades since the map was originally commissioned as part of the Arctic Waste response program of the Department of Indian and Northern Affairs (DIAND), only one site – Iqaluit's "Upper Base," a 1950s-era DEW Line radar station – has been fully remediated (Eno 2003). The other waste dumps remain: a more than half-a-century-old metal dump is located in the nearby territorial park, right next to the city's most popular campsite; three other waste sites, including the city's dump, are located at Causeway, the city's main launch point for those going out on the land to hunt and/or camp; and two others (a metal dump, and the contaminated North 40 site) that are centrally located between the new expanded airport, a college residence, and the territorial penitentiary. These sites, as the councillor who wrote the article explained, "are a threat to that [Inuit] way of life" (Dobbin 2013, 1) because of their desorption into local "country" foods, such as berries and aquatic wildlife (i.e., seal, narwhal, and shellfish). These active dumps are more than colonial reminders; they are symptoms of ongoing settler colonialism.

## AND THEN SOME

The MSW openly dumped in Arctic communities is a small fraction of Canada's Arctic waste portfolio. The construction of the Canol pipeline in 1942, for instance, saw some 40,000 military personnel and civilian workers move into the North in order to secure a pipeline from Norman Wells, Northwest Territories, to Whitehorse, Yukon. Operational for just one year, the pipeline closed because more oil was spilled on the land than was transported (Perić 2015). When workers left this military-industrial enterprise, they abandoned hundreds of trucks, graders, and construction equipment, as well as some 60,476 barrels of oil in the pipe, and some 108,857 barrels that are presumed to have spilled into the landscape (*Up Here* 2014). Also constructed in 1942 in the wake of the attack on Pearl Harbour, the Alaskan Highway would eventually connect the contiguous United States-to-Alaska border through Canada's British Columbia and Yukon. Although celebrated for having taken less than a year to build, the highway was mired with technological problems caused by permafrost – challenges that still cause rerouting. The project moved thousands of tonnes of construction equipment north, and the work crews dubbed the road the "oil can highway" because of the sheer number of oil cans and fuel drums they discarded along the road. The Arctic's waste, moreover, has also accumulated "exorbitantly" (see N. Clark 2005): in January 1978, the Soviet satellite Cosmos 954 exploded in the atmosphere over the Northwest Territories, spreading some sixty-five kilograms of fissionable uranium-235 over an area of 124,000 square kilometres (Heaps 1978), providing another dramatic illustration – like the plastics and persistent organic pollutants (POPs) drifting northwards and accumulating in the Arctic from industrialized southern regions – that waste does not respect national boundaries.

Moreover, there are approximately – no one knows the exact figure – 27,000 abandoned or "orphaned" mines in Canada, most of which are in Canada's northern regions. And as chapter 1 briefly described, the Giant Mine, located on the Ingraham Trail close to Yellowknife, was abandoned in 2005, leaving responsibility to the Ministry of

Indian Affairs and Northern Development and taxpayers for the cost and cleanup of some one-hundred on-site buildings, eight open pits, contaminated soils, and waste rock around the mine, including some 237,000 tons of arsenic trioxide dust (Sandlos and Keeling 2012).

The "grasshopper effect" is a term used to explain how POPs from all over the world end up in polar regions. Many POPs are industrial waste by-products that are transported to the Arctic via air currents from southern communities. Contaminants evaporate in warm temperatures and condense in cold climates, where they accumulate on the land and in country food. As a result, women who eat country food have higher contaminant loads in their breast milk than those who do not (Kafarowski 2004; see also Cone 2005).

Disposal of sewage and grey water at sea is regulated in all Canadian waters except for the Arctic Ocean, where "any ship and any person on a ship may deposit in the arctic waters such sewage as may be generated" (*Arctic Shipping Pollution Prevention Regulations*, C.R.C., c. 353, s. 28). The thousands of vessels travelling the Arctic since 1990 are mainly for tourism and research, and federal government military support vessels (Pizzolato et al. 2014). A recent study reveals that concentrations of microplastics (plastics debris that is less than five millimetres in diameter) in high Arctic sea ice is over two orders of magnitude greater than what is found in all other ocean surface waters, including in the so-called the Great Pacific Garbage Patch (Obbard et al. 2014). The research concludes that the Arctic is a "global sink" (ibid., 318) for microplastic debris – one that will result in a substantial release of plastic particles into the ocean upon human-induced sea ice melt.

What is known as the Highway of the Atom (see van Wyck 2010) refers to a route travelled by uranium that was mined by Sahtú Dené on the shores of Great Bear Lake, over land to Port Hope and then to the United States for use in the Manhattan Project, where it was eventually dropped on Hiroshima and Nagasaki during the Second World War. In the 1930s and 1940s, the Sahtú Dené were employed by the Eldorado Gold Mines to mine uranium ore without knowing of its effect on their health and their environment (including tons of radium tailings dumped into the lake, and through many kilometres of ground). Generations later, they are still living with its aftereffects: depression, addiction, cancers, and mutating (and dying) flora and fauna.

## THE DEW LINE AND CANADA'S MILITARY WASTE LEGACY

According to the Department of National Defence's glossy publication *Defence Environmental Strategy* (n.d.), the Canadian military manages approximately 2.25 million hectares of land in Canada. On this land there are defence establishments and infrastructure, military equipment, electronics, fleets, and ammunition and explosives. There are also various hazardous materials: petroleum, and oils and lubricants; radioactive material and ozone-depleting substances; solid waste; and wastewater and effluents. Some of this military waste is a by-product of contemporary training exercises and military operations, while other waste reflects our military's history. Beyond this, Canadians know very little about either the extent or severity of military waste in Canada, nor the Canadian military's waste legacies and waste production in other countries (as a by-product of its operations, peaceful or aggressive).

A more exacting list of the kinds and amounts of materials is difficult to acquire, since much of this information is classified. We typically have to wait many years for classified documents to be released, military territory to be sold, or, in a number of cases, for waste disasters to occur, which then bring military waste issues to the public's attention. For example, we now know that on 21 February 1946, press and other media were invited to board the *General Drury* in order to witness the Canadian military dump into the ocean 10,982 forty-five-gallon barrels of mustard gas, 180 miles south-east of Halifax, at a depth of some 1,350 fathoms. The mustard gas was manufactured by Stormont Chemicals, in a secret warfare factory in Cornwall, Ontario (Kehoe 2002). The United Kingdom, the United States, and other countries dumped mustard gas as well as munitions and other contaminated materials off their coasts.[6] Indeed, the Canadian Air Force's "Military Top Priority One Classification" includes twenty-eight sites "known to have reactive and viable chemical and biological weapons or nuclear weapons" (Kehoe 2002, Petition No. 50A). The location of ten of these sites appears on local fishing charts; the location – and what each site contains – of the remaining eighteen remain unknown, even to the military. And even when the military does know the location, "security and safety considerations could preclude the release of the exact locations due to the materials that may be present at these sites," stated John McCallum, minister of National Defence (ibid.). Greatly understating the problem, Wendell King, former colonel in the US Army, wrote, "Our legacy of environmental protection and military installations is not the most exemplary part of our military history" (King 2001, x).

The military waste site for which we have the most data is the DEW Line. It was constructed to protect North America from a Soviet invasion via the Arctic. Sixty-three stations stretching across Alaska, Canada's Arctic, Greenland, and Iceland left behind a legacy of exposing Arctic communities to waste contamination, pollution, and abandoned infrastructure. In the wake of extensive remediation, the DEW Line's waste endures as a pivotal artefact of complex and interwoven discourses of sovereignty, security, resource exploitation, and Inuit self-determination.

In the wake of the Second World War, efforts to secure the United States and Canada primarily focused on the military protection of the Arctic. As tensions between the United States and the former Soviet Union grew, US Republicans, tipped off by scientists at the Massachusetts Institute of Technology (MIT) that the US was vulnerable to northern Soviet attack, began to envision a technologically advanced military response. The demand that eventuated was both simple and direct: "the Soviet Union was the enemy, war a possibility, bombers ... the threat, the polar corridor the route" (Eayrs 1972, 335). Any objections concerning the high costs and ineffectiveness of the plan ended when news surfaced that the USSR had successfully tested a hydrogen bomb in 1953. In about five years – between 1952 and 1957 – scientists, the military, and corporate America worked in tandem to build the DEW Line. United States Secretary of Defence Robert A. Lovett – whom G. William Domhoff dubbed the "Cold War architect" – contracted the private corporations Western Electric and Bell Technologies, who worked with the US and Canadian militaries to build a number of radar stations – the DEW Line – that would stretch the width of the Arctic, from Greenland, through Canada, and on to Alaska (Pigott 2011).

When it was completed, the DEW Line provided a two-hour warning of Soviet missiles directed at the US. As *National Geographic* correspondent Howard La Fay enthusiastically wrote in 1958:

In the event of an enemy attack across the polar ice, the DEW Line will flash instant warning to the joint Canadian-U.S. combat operations center at Colorado Springs, Colorado. The time thus gained could spell the difference between national life and death for Canada and the United States ... the civilian population would take cover. (1958, 129)

While the rationale for building the DEW Line emphasized the importance of advance warning to protect American and Canadian civilian populations, Heather Myers and Don Munton (2000) argue that the real intent was to alert the Strategic Air Command (SAC) of the United States Air Force (USAF) so that it could launch a counter-attack. This technological innovation proved short-lived, however, when the Soviets launched Sputnik in 1957, demonstrating that new ballistic missiles would allow far shorter advance warning. DEW Line sites began to close in 1963, just six years after their completion (the last one closed in 1993) (Loock 2014). Twenty-one of the sites were abandoned, becoming the responsibility of DIAND to remediate (Bennet et al. 2015). Twenty-one further sites were redeveloped into the new North Warning System, requiring the movement of more equipment, more infrastructure, and more military to the Arctic (Loock 2014). Nevertheless, the DEW Line's waste legacy continues: the asbestos, polychlorinated biphenyls (PCBs), and other contaminants in the soil, water, and atmosphere that have not been shipped back down south remain in the Arctic.

The DEW Line construction cost approximately $7 billion and involved 23,000 workers and 45,000 planeloads of radar equipment and machinery: 46,000 tons of steel, 75 million gallons of fuel, 22,000 tons of food, and 12 acres of bed sheets (Capozza 2002, 14; Ducharme 2004). In a rather understated fashion, Kevin O'Reilly, research director for the Canadian Arctic Resources Committee, remarked that there is "no question that the DEW Line dramatically changed the Arctic – economically, socially – but also environmentally" (in Capozza 2002, 16). When Canada forged the agreement with the United States to build the DEW Line, Canadian officials attempted to mitigate their "younger sibling" position with detailed directives. The "Canada–United States Establishment of a Distant Early Warning System: Agreement between Canada and the United States of America" (Canada–United States 1955)

Figure 5.2 | Garbage at Cape Dyer, Nunavut DEW Line site.

explicitly specified Canadian sovereignty, that the infrastructure would be built by mutual agreement between both countries, that the plans would need Canadian approval, that Canada would have the right to inspect all construction work, that Canadian contractors would have "equal consideration" and that Canadian labour would be preferentially considered, that any scientific data collected during the construction and maintenance of the Line would be shared between both countries, that Canada would play a decisive role in the operation of the sites, and that affected "Canadian Eskimos" would enjoy limited protections.

Yet, when Canadian representatives visited the DEW Line sites, they found that most were only flying the US flag. Moreover, all Canadian personnel working on the line had to, by US policy, have a US security check, and be supervised by an American. After much political wrangling with American officials over several years, with further stalls by the US Congress, the US eventually agreed to

Figure 5.3 | "Goody bags" of PCB-laced soil await shipment to southern incinerators from Cape Dyer DEW site.

pay $100 million from 1995 to 1999 towards remediation – and this was in the form of credit towards the Canadian government buying US military equipment (essentially spare parts). In the *ex gratia* agreement, the United States expressed that "it is the view of the United States government that it has no legal obligation under current US and international law to reimburse the costs of environmental cleanup," and that the agreement was a "full and final settlement of all claims for costs of environmental cleanup at the four installations" (referring to the DEW Line, the USAF base at Goose Bay, the US base at Argentia in Newfoundland, and the failed Haines–Fairbanks pipeline). In all, the US has contributed about one-fifth of the cost of remediating one-quarter of the DEW Line sites.

Of the total sixty-three DEW Line sites, forty-two are in Canada, and twenty-one of these Canadian sites have (as of 2014) been remediated through a series of agreements involving Aboriginal Affairs and Northern Development Canada (AANDC), the Department of National Defence (DND), and Nunavut Tunngavik through a series of four agreements in 1996, 1998, 2001, and 2005 (Loock 2014). In 2014, The DND announced it had completed its remediation program. When the DEW Line was officially abandoned, most of the American and southern Canadian workers (who made up almost all of those employed to build and maintain the line) returned south (Ducharme 2004). They left in their wake abandoned sites contaminated with various toxic chemicals that have had to be removed – in some cases, square centimetre by square centimetre – to southern Canada for treatment (typically burial in a landfill or incineration), at an estimated cost of over $500 million.

The various DEW Line remediation projects have uncovered an extensive list of contaminated and uncontaminated waste: waste oil; PCB transformers and capacitors, asbestos, sewage, lead-based paints, radioactive tubes, scrap metal, radar components, fuel barrels, lime, antifreeze, wood, aviation fuel, sulfamic acid, cathode ray tubes and screens, filtron tubes, oscillators, meters, copper wire, transmission fluid, 1-1-1-trichloroethane, PBX telephone equipment, mercury vapour rectifier tubes, paint thinners, batteries, chlorinated hydrocarbons, corrosion inhibitors, lye, corrosives, paper, plastic, solvents, dynamite, RF interference filters, generators, scopes, vehicles, and rubber fuel bladders (Environmental Sciences Group and UMA Engineering 1995). Over 30 tons of PCBs were found up to 15 kilometres away from various sites.

Scientific reports detail the scale of the remediation. At the DEW Line site at Resolution Island, which Scott Mitchell (director of the DIAND contaminated sites remediation program) estimates represented a third of the total DEW Line remediation costs, scientists found that over 8,000 kilograms of pure PCBs (Aroclor 1260) had been abandoned after the site closed (Kalinovich et al. 2008). These PCBs had leaked out through a valley, descended cliffs, and moved into Frobisher Bay, from which local Inuit fish. The fuel and oil that had also been openly dumped, forged a path for the PCBs, facilitating their migration over the land and into the sea. Describing the scene scientists found at Resolution Island, Canadian hazardous waste specialist Robert Eno said: "Looking at what you'd found there, you'd think that Americans took big hoses and sprayed PCB liquid all over the site" (in Capozza 2002, 15). PCBs are a known carcinogen, increasing the incidence of cancer, bacterial infections, liver lesions, and genetic defects when exposed. PCBs have shown up in polar bears, foxes, voles, trout, and other country food upon which Inuit depend (Danon-Schaffer 2015). This site alone cost $64.75 million to remediate and the operation involved 595 people (Kalinovich et al. 2008).[7]

Oil spills in Hooper Bay, Cape Romanzof, and Point Hope amount to some 80,000 gallons (303 m³) of petroleum leaked into the environment. Some of the abandoned barrels contained petroleum products (fuel, lubricants); antifreeze (glycols); degreasers (halogenated aliphatics); and cadmium, chromium, lead, and chlorine, as well as the staggering amounts of PCBs. But scientists found the highest levels of PCBs at the Sarcpa Lake site, as well as solvents, mercury, and petroleum products; buildings and other infrastructure; and abandoned barrels, sewage, and other debris (Poland, Mitchell, and Rutter 2001). At the Iqaluit and other DEW Line and Pinetree Line sites, scientists had to remove not only the buildings, which had asbestos-clad piping, but also thousands of barrels abandoned from the DEW Line project, including those left on the site from more recent industrial activities.[8]

Waste remediation in the Arctic often requires physically moving contaminated waste to southern disposal facilities – facilities that are completely absent in the Arctic due to their very high associated costs, and physical challenges such as permafrost (see, for instance, Thomassin-Lacroix 2015). At the Iqaluit site, not only did the buildings require demolishing, but the concrete foundations also had to be removed and shipped south for disposal, because the PCBs had entirely penetrated the concrete to between 50 and 70 centimetres (according to the 2005 DEW Line Cleanup Protocol [DLCU], PCBs at more than 50 ppm must be shipped to a licenced disposal facility) – using non-renewable fossil fuels in the transportation. The PCB-contaminated drums found at the DEW Line site in Iqaluit had to be flown from Iqaluit to Yellowknife, and then land-transported to a PCB incineration facility in southern Canada (Poland, Mitchell, and Rutter 2001). A further 307 metres cubed of soil contaminated at the DCC (DEW Line Cleanup Criteria) Tier II level (5–50 ppm) was placed in specially designed fabric boxes and shipped to a waste disposal site near Montreal. Engineers had the extra challenge of designing viable landfills that could be excavated in hard rock and permafrost. As well, technologies used in the south such as incineration, thermal desorption, and solvent extraction prove to be unviable in the

Arctic, as they are too expensive, require very large amounts of fuel (which would need to be shipped north), produce residues that must be shipped south for disposal, or incompletely deal with the contamination. Describing the Resolution Island cleanup, scientists noted the unique challenges of waste remediation in the North: "The magnitude of the PCB contamination is very large, the terrain mountainous, the site extremely remote, the climate is particularly harsh and polar bears are regular visitors" (Poland, Mitchell and Rutter 2001, 96).

And although the term "cleanup" suggests that scientists have been able to return the sites to their pre-construction condition, remediation has been the more attainable goal. The term "remediation" has an important caveat. In a report discussing the lengthy assessment of the feasibility of cleaning up the DEW Line sites, the Environmental Sciences Group (ESG) concedes, "The Protocol recognizes that this restoration will not return the environment to a pristine state, but will at least remove most barriers to long-term natural reclamation" (1995, 30). And, further, as Heather Ducharme (2004) observes, whether the terms "cleanup" or "remediation" are used, both actually involve further material development of landfills and burns. Even the assessment process has involved further environmental disturbance: the "environmental sampling to survey the damage has involved some 4,000 soil/sediment/water, 1,600 plant, and 500 marine/animal tissue samples" (ibid., 15). Moreover, much of the material brought to the Arctic – if it is below the protocol's threshold for contamination – has been left there. And the vast quantities of material (soil, barrels, and so on) that are so contaminated that they must be moved to the South, are then not cleaned up but rather buried in southern landfills, or incinerated – a process producing highly toxic fly ash that must be stored in specially designed facilities (Rowe 2012). So far, then, our best response is to either leave hazardous and nonhazardous waste where it is, try to isolate hazardous waste, or transport the hazardous waste somewhere else, where it will be either landfilled or incinerated.

## CONCLUSIONS

The ESG report repeatedly suggests that the DEW Line was constructed at a time when less was understood about contamination and its effects, and that it is for this reason that such widespread and damaging contamination occurred: "It became apparent that many substances, including PCBs, had been 'poured down the drain' or in areas adjacent to buildings" (ESG 1995, 16). But the DEW Line's waste – its remainder and legacy – endures within a much larger waste landscape, one that stands to significantly expand and become more complex as the focus on oil, gas, and mineral resource exploitation greatly intensifies, and issues of sovereignty and security remain paramount. The issue is also made more challenging by the fact that our knowledge will always be in a state of flux as we increase our understanding of the harmful effects of waste contamination. Thus, we must be cautious about justifying environmental degradation in the name of partial knowledge, since this is our society's constant state.

As such, Inuit leaders are well justified in bringing to the forefront the precautionary principle as an organizing tenet for all present and future resource exploitation. The waste legacy of the DEW Line strongly suggests that a "develop now; pay later" approach incurs profound environmental and human health consequences, particularly for Inuit. Much will depend upon the Canadian military's emerging role in environmental protection: to date, the military's focus has been on traditional sovereignty and security concerns that increasingly involve providing material, infrastructure, and personnel support for resource exploitation. Insofar as the Arctic's waste legacy may be used in considering future Arctic waste scenarios, Canada's Northern Strategy must redefine the responsibilities of military and industry in terms of waste prevention and remediation as a critical part of Canadian sovereignty, security, and environmental protection – the producer must pay.

CHAPTER SIX

# Wasting Animals

## INTRODUCTION

As the unintended, and often unacknowledged, fallout of capitalism and settler colonialism, humans have developed sophisticated technologies to squirrel away our discards: as the previous chapters have detailed, waste is buried, burned, gasified, thrown into the ocean, or otherwise kept out of sight and out of mind. Despite efforts to disgorge ourselves of waste, millions of people live with, and on, consumption's cast-offs (M. Davis 2007). Additionally, an undocumented number of "trash animals" – gulls, ravens, pigeons, raccoons, rats, mice, dogs, polar bears, and so on – eat, defecate on, play games with, have sex on, and otherwise live out their lives in our dumpsites (Nagy and Johnson 2013). As this chapter will explore, culturally sanctioned and publicly funded modern facilities in southern parts of Canada practice diverse methods of "vermin control," legitimated within discourses of public hygiene and safety (ibid., 3–8) – practices that for the most part do not occur at northern dump sites.

As the previous two chapters have detailed, waste and wasting in the Arctic exist within a complex set of historically embedded and contemporaneously contested settler colonial regulations, policies, and formal and informal practices. In Iqaluit, Nunavut's capital, ravens rest on open dumps, and polar bears who have become sick from exposure to, and ingestion of, waste may be killed out of respect and safety concerns. This chapter reflects upon why animals are "managed" at modern landfill sites across southern Canada and left to scavenge on open dumpsites in Canada's Arctic.

My graduate student, Alex Zahara, and I had many conversations about animals in Iqaluit and other Arctic communities. I have long been fascinated by ravens, both the role they play as central figures in Inuit cosmology (and Indigenous cosmologies generally) and the rich scientific studies documenting raven intelligence, ingenuity, and playfulness. Our conversations were brought into sharp relief when we were driven past a small lake in Iqaluit where local residents say that the bodies of sled dogs killed by the RCMP were dumped, or when we were standing at the base of West 40, watching ravens casually dive onto the mountain of garbage.

The way animals are variously managed, both Alex and I argue, is not simply a matter of modern-versus-outdated waste disposal technologies and practices – although this is a central way in which waste issues in the Arctic are framed by government officials and the media. In this chapter, I explore the ways in which historically and culturally embedded practices inform particular relationships with the inhuman.[1] Canada's Arctic is a site where differing cosmologies variously collide, intertwine, operate in parallel, or often speak past each other in ways that often marginalize Inuit ways of knowing and being with animals and landscape.

This chapter examines how encounters with the inhuman have been, and continue to be, integral facets of the Canadian settler colonial project. Building on the previous two chapters, I will argue that the settler colonial legacy of managing Inuit, animals, and the Arctic landscape is a direct outcome of the neoliberal capitalist, settler colonial venture that forefronts Canadian state sovereignty. I follow this mapping of capitalist venture and neocolonial governance with a discussion of the burgeoning interest in those inhuman creatures who survive through relations with human debris. This literature points to the complex and often contradictory Western understandings of animals as "companion species," who are variously cherished, pampered, used as labour, abused, discarded, and killed (Haraway 2008).

Our attention then turns to two particular animals – ravens and sled dogs – whose iconic presence in the Arctic exemplifies the complex and often contradictory understandings of the inhuman within this particular settler colonial landscape. Ravens and sled dogs feature in Inuit cosmology, hunting practices, and culture, and both have endured – however tentatively – a rapidly and profoundly changing status in Canada's Arctic. This change has occurred, in part, because waste and its associated inhuman creatures are "othered" within settler colonial governance practices. Across Nunavut, this has contributed to the changing image of ravens from Creator to pests scavenging from open dumpsites, and to the killing of thousands of Inuit sled dogs, whose deaths have forever changed the way Inuit experience human/nature relations and landscape, as well as labour practices and self-determination. Using the Canadian Arctic as a case study, we explore the ways in which waste and our associations with waste inform the settler colonial present. We argue that inheriting waste is more complicated than just a relay of potentially indestructible waste materials from past to present to future: through waste, we bequeath a set of politically, historically, and materially constituted relations, structures, norms, and practices with which future generations must engage.

## "TRASH ANIMALS" AND THE ARCTIC

As Donna Haraway argues in her pathbreaking work, capitalism's technoculture structures particular relationships with the inhuman. From agility training, to medical and hygiene practices, to the selection of financially lucrative genes, the ways in which we encounter our inhuman companions characterize them as "lively capital" (2008, 45). Even shepherding and livestock dogs, whose companionship – both as labourers and as family members – has historically been requisite for the survival of many humans, developed over centuries within the context of both nomadic and sedentary human livelihoods.

Some of our urban companion species have adapted to our capitalist lifestyles not at the point of production and consumption, but at the point of disposal, where there is a normative shift in our encounters with animals. Combining waste and animal studies, a number of scholars are examining the treatment of human and inhuman urban "scavengers." As was examined in chapter 2, "trash is not just the material stuff we throw away, but a classification that defines for us the ways we understand and act toward certain inanimate and animate objects" (Nagy and Johnson 2006, 5). "Trash animals," as Nagy and Johnson argue, are despised, feared, and mocked – they have become a "disgusting 'other' in our anthropocentric fantasies of existence" (Malamud 2013, ix).

Settler colonial studies, waste studies, and critical animal studies explore the ways in which humans and animals are depicted through their associations with garbage. Postcolonial theorists, for instance, have examined how waste contributes to the othering of marginalized groups. As chapter 4 details, settler colonialism in Canada and elsewhere has long associated Indigeneity with waste, not least as a way of justifying colonial structures and practices of subjugation and violence. This often occurred under the pretence of safety or civility – or what Marie Lathers refers to as "management of the abject" (Lathers 2006, 419). Warwick Anderson's discussion of fecal waste in the Philippines, for example, notes that American colonizers assumed responsibility for ending Philippine peoples' "promiscuous defecation" (W. Anderson 1995, 642). The practice of teaching colonized people Western notions of hygiene is rooted, as Kay Anderson argues, in the Judeo-Christian idea of humans transcending "the so-called 'bestial' within" (2000, 4). Unlike animals – which, within an anthropocentric epistemology, are confined to their biology – humans, with their capacity for reason, should "overcome nature" by suppressing their bodily functions and otherwise hiding their wastes. More than anything, the problem with "dirty" Filipino bodies was that they prevented American colonizers from eschewing the nature of their *own* bodies; the settlers were forced to recognize that they themselves were privy to the abject in the form of their own shit. In the colonial tradition that considers Indigenous wastes as "uncivilized" and "a problem to be solved," the association of Inuit with waste both prompted and justified many assimilative government policies and practices (Lathers 2006, 419). As chapter 4 examines, in the mid-twentieth century, teaching Inuit how to properly interact with waste – to excrete indoors, to avoid dirt, to eat using dishes and utensils – was a priority for the Canadian federal government (Harrison 1964, 64, 100–6). For instance, when the DEW Line radar stations were being constructed in the late 1950s by the Canadian military (examined in chapter 4), nomadic Inuit families camped near DEW Line dumps (which were created by the American and Canadian militaries and industries) to scavenge for discarded food and other reusable materials. Inuit caught scavenging at the dump were described by the RCMP as "bums and scroungers" (QIA 2013e, 60). RCMP and DEW Line operators, who considered the Inuit practice of scavenging offensive, responded by burning their own unused food rather than sharing it with food-insecure Inuit families.

Indigenous studies scholars in the field of critical animal studies have examined how Western understandings of nature mix uneasily with Indigenous cosmologies. Within Inuit cosmology, the environment has *anirniq* – "breath" or "spirits" (Qitsualik 2013, 29). According to this cosmology, Inuit hunters must follow a series of rules and modes of conduct made known to them through shamans, and passed on through generations of practice and storytelling (Henri 2012, 169–70). If these rules are followed, eventually an animal will allow itself to be killed, its *anirniq* will pass on to another animal, and nature will be respected (ibid.). Moreover, Inuit cosmology recognizes that, while Inuit lives are entirely dependent on the land (*nuna*) for survival, *nuna* – along with the life and breath that it supports – continues to exist with or without Inuit.[2] Inuit cosmology also implicitly recognizes that one can never fully know nature. For Inuit, being confident in one's relationship with the landscape is considered hubristic, dangerous, and counter to the understanding of *nalunaktuq* – nature's unpredictability (Qitsualik 2013, 27). As Emilie Cameron and others point out, "uncertainty, unpredictability, and change" are foundational to Inuit understandings of, and relationships with, nature (Cameron, Mearns, and McGrath 2015, 5).

And so, in the early days of settler colonialism, what was for Inuit simply "a default to universal order" was for settlers unadulterated "chaos" (Qitsualik 2013, 27). When Christian missionaries and explorers arrived in the Arctic, they brought with them an entirely different understanding of nature. The reciprocity between Inuit and the inhuman embedded within Inuit traditions was all but entirely superseded by southern Canadians, whose religion, culture, and identity were maintained through what they saw as their own independence from, and superiority over, nature (ibid., 32). In the first instance, missionaries taught Inuit that only humans have souls, and that God decreed humans' pre-eminence: "Let us make man in our image, after our likeness: and let them have dominion over the fish of the sea, and over the fowl of the air, and over the cattle, and over all the earth, and over every creeping thing that creepeth upon the earth" (Genesis 1:26, King James version). Colonial settlers considered it their right and responsibility to dominate the environment; it was to be made "disciplined," "tidy," and "orderly" (Qitsualik 2013, 27).

In Nunavut today, disparate ideas concerning the relationship between humans, animals, and the environment come to a head, often in controversies over hunting.[3] These controversies are predicated on fundamentally different understandings of relations between humans and animals. For example, in 2007, Matt Rice, the anti-sealing campaign coordinator of People for the Ethical Treatment of Animals (PETA), wrote to Iqaluit City Council asking the city to lower its Canadian flag to half-mast as a way of mourning the seals killed in the community's annual seal hunt. When city council rejected the proposal, he told *Nunatsiaq News* that PETA considers the sale of seal pelts "a matter of waste and extreme cruelty" (Windeyer 2007). Similarly, the European Union has banned the commercial use and import of all seal skins and seal products, stating concerns about "animal welfare" (European Commission 2014). These events follow the Greenpeace-supported ban on sealing in 1976, which contributed to mass Inuit food insecurity and poverty in the 1980s and 1990s. More recently, comedian and talk show host Ellen DeGeneres urged her fans to sign a petition condemning the seal hunt and to tell Canadians (i.e., Inuit) that "killing innocent animals is wrong" (Ellen TV 2011). Inuit responded with a "sealfie" movement, taking pictures of themselves next to dead seals, seal fur, or seal meat and sending these pictures via social media to the *Ellen* show (Bowman 2014). When Inuk throat singer and activist Tanya Tagaq tagged a picture of her baby next to a dead seal using the "sealfie" hashtag, she received death threats from animal rights activists. For Tagaq, the photo was meant to convey an acknowledgment of the relationship between Inuit and seal. As she explained, "One of the traditions is to melt snow in your mouth and then put it into the seal's mouth so their spirit isn't thirsty in the afterlife ... I put my baby there to show how peaceful it can be and how much you can respect the animal" (Freeman 2015). According to Tagaq and other Inuit, attempts to prevent Inuit from hunting are "a mini version of colonialism" faced by Inuit today (Tagaq 2014). The documentary film *Angry Inuk*, directed and narrated by Alethea Arnaquq-Baril, details the Inuit struggle to reverse the European Union's ban on seal products. Rejecting the EU's argument that an exemption should only be granted for the purposes of Inuit (and other Indigenous peoples) directly consuming the food they derive from their own hunting activities, Arnaquq-Baril argues that Inuit heavily depend on selling seal products as part of the profit-driven commercial hunting business that settler colonialism forced into Inuit culture in the first place. That is, settler colonialism created the dependency of Inuit on a capitalist market economy (through mechanisms such as sled dog killing, as we will shortly see) and is now punishing Inuit for this violently created and enforced dependence.

Yet contrast Tagaq's explanation of Inuit "becoming-with" seals through the practice of hunting with non-Inuit anthropologist Rick De Vos's description of an interaction with Greenlandic Inuit: "My comments to locals [Inuit] regarding my desire to see narwhal, beluga and walrus, generally led to exclamations and responses praising the deliciousness of these animals as food, with little understanding of my wanting to spend time with these animals without hunting and eating them" (De Vos 2013, 286).[4] The history of settler colonialism embedded in anti-hunting campaigns exemplifies a paradox of modern Arctic sovereignty that is bequeathed through particular settler colonial traditions (Qitsualik 2013, 32). Tagaq succinctly describes Inuit cosmology that does not separate animals, humans, and nature: "We're the same. We're flesh, we're meat, we're so stupid to think that we're not ... A wolf is not evil when it hunts a caribou" (CBC Radio Q 2014). Within Inuit cosmology, hunting and respecting animals are not mutually exclusive, and food garnered from hunting is neither dirty nor wasteful.

In the following section, I will take a closer look at two familiar trash animals – ravens and gulls – that are differently "managed" in Arctic dumps and southern landfills. Their management is informed by very different understandings of humans, trash, animals, and nature. I will examine traditional Inuit relationships with animals, not as legends or myths (and therefore as points of erasure), but as a way to more accurately represent how settler colonization was (and still is) experienced in Nunavut communities (Watts 2013). From this analysis, we move to a discussion of sled dogs, and a particularly traumatic event in the history of colonial settlement in Canada's Arctic, made possible by the reconfiguration of these dogs as a hazardous threat to be resolved through confinement, and extermination/wasting.

### GULL, TULUGAQ

From late April through to the end of May, the critter population in Yellowknife increases substantially, as migratory birds, including gulls, ravens, raptors, sandhill cranes, and magpies, flock to the Great Slave Lake region to construct nests, lay and incubate eggs, care for their offspring, and then help their young leave the nest in a process ornithologists and bird enthusiasts know as "fledging." Yellowknife is the capital of and largest community in Canada's Northwest Ter-

Figure 6.1 | Hawk used at a landfill, servicing Montreal, Quebec.

ritories, and as small as it is relative to southern Canadian urban centres (the population hovers under 20,000 people), Yellowknife shares with its urban southern neighbours a particular approach to the 256 or so species of birds in the region, and the 30,000 tons of human waste the city produces (Beacon Environmental 2008, 1). Yellowknife and southern Canadian communities, like those in North America generally, relate to the birds frequenting their landfills as largely "wildlife hazards" and "vermin" that require industrial management (ibid., 28). Move to the east some two thousand miles from Yellowknife to Iqaluit, and the scene is quite different. Hundreds, if not thousands, of ravens casually circle the town's open dump, swooping leisurely to land on fresh piles of discarded food that the steady flow of trucks deposit on the colossal dump. The ravens are not in any hurry to grab the wasted food and fly off; they take their time. This is, in some ways, their dump, and no one attempts to scare them off.

Southern Canadian landfills typically adopt a "zero tolerance" approach to trash animals. Gulls are most certainly the lowly, senseless, and reckless underclass of the modern landfill, or "bird buffets" as they are colloquially called. Landfills are a primary food source for the gull: in under 15 minutes at a landfill, gulls are able to satisfy their daily nutritional requirements (Department of Environmental Protection 1998, 4). According to a Massachusetts Department of Environmental Protection report, the goal of any landfill is not only to completely prevent gulls from feeding but also to "eliminate or reduce the suitability and attractiveness of the facility for other gull activities, such as resting, roosting, or loafing" (ibid., 4). Gulls and other birds like to "loaf" – communally rest, bathe, drink, or preen – on landfills because the expansive and relatively flat space provides good visibility for spotting predators (Beacon Environmental 2008, 5).

On my many trips to landfills, I have found it difficult to witness much bird loafing on landfills in southern Canadian communities. Responding to community complaints about the noise birds make and the excrement they leave on the roofs of houses, as well as the risks birds pose to nearby aviation – where birds risk damaging aircraft and potentially injuring passengers – landfill operators focus on what they euphemistically call "wildlife management." Landfill operators do not like the attention scavenging birds bring to landfill sites: they prefer trash to disappear from people's minds once it leaves their curbside. Thus, operators have introduced a cacophony of management techniques, including canons, air-operated human effigies, scarecrows, chemical repellents that poison the land, distress calls, pyrotechnics (including bangers, screamers, and flaming whistles), tape ribbons and other shiny objects, helium-filled "evil eye" balloons, decoys, collecting and oiling eggs (which kills the developing birds through suffocation), and even displayed bird carcasses (both real and facsimiles) (ibid., 29–30; Cook et al. 2008; R. Johnson 2009; Parrilla 2012). But problems abound: gulls turn out to be smart, and quickly figure out that the scare tactics are distractions. Moreover, as landfill operators note, pyrotechnics and other strategies "will alarm and surprise some [human] landfill customers, sometimes with very emotional effects" (R. Johnson 2009). Faced with smart, adaptive birds and skittish people, operators have recently introduced falcons and hawks to patrol the landfill landscape. These birds of prey have become part of WM's big business. They, according to hawk handlers, are allowed to eat any gull they catch, although this is not particularly good for the hawks because the gulls may carry contaminants through their contact with leachate, which is the putrescible and organic landfill material transported by water. Female falcons are typically used because they are bigger than their male counterparts, and more aggressive: posing a potentially added harm to future generations of falcons when they produce offspring from bodies contaminated from ingesting human waste.

In disconcertingly earnest statements, El Sobrante landfill spokesperson Miriam Cardenas exuberantly exclaimed, "We are using nature to control nature. It's the most effective method" (Parrilla 2012) and falconer Jorge Herrera described his work as "nature taking care of itself" (ibid.) – while the falcons he uses sit on perches on the back of his truck with tracking devices attached to their ankles. Here are birds of prey captured and implicated into a complex assemblage of bird-waste-human-landscape to live out their lives terrorizing other birds so that people are not disturbed by bird calls, rooftops can remain unsoiled, and airport runways can be expanded. Enlisting hawks to distress and kill gulls makes good entrepreneurial sense: to wit, the hawks have silenced community complaints. The hawks, then, become part of the geoengineering architecture – not dissimilar to the covering over of landfills and their transformation into suburban sprawl – that encourages people to forget about waste beyond their curbside.[5] Landfills in southern

Canada tend to be sited away from upper- and middle-class communities and cordoned off behind high fences, to be managed out of sight. As the introduction of this book details, waste is something we do not want to remember, or be remembered for, and WM corporations profitably remove our waste from consciousness (van Wyck 2012).

By contrast, waste in many of Canada's Arctic communities is left there, on the land, in highly visible dumps; raw, uncompromising, and unapologetic. These Arctic landscapes are not covered over and they are not out of mind. Whereas southern landfills are populated with gulls, *tulugaq* (ravens) are the most common birds in northern communities, and remain in the Arctic throughout the winter. According to Inuit creation narratives, *tulugaq* made the world and the waters with the beat of his wings (Blake 2001). *Tulugaq* the Trickster is respected for his resilience, intelligence, and sociability. *Tulugaq* teaches children how to live in community, and newborn Inuit boys are clothed in raven skin to help them become successful hunters (Polarlife n.d.). *Tulugaq* follow polar bears and scavenge leftover carcasses, and Inuit mimic the raven's "caw" to attract polar bears while hunting. *Tulugaq* also call wolves to dead animals so they will make the carcasses more accessible to the birds. As I have watched ravens swoop over various dumpsites in Canada's Arctic, I have wondered if they now call humans to these dumpsites to leave fresh trash.

In Inuit narratives, ravens possess the ability to transmute – presently, it seems, into garbage pickers. *Tulugaq* have followed the transition of Inuit from a nomadic lifestyle, in which *tulugaq* assisted hunters in their search for food, to sedentary community living, whereby food is found conveniently left on the landscape at the community dump for ready picking. As Kerry McCluskey observes, "instead of dipping their wings to point to a polar bear, ravens are now more likely to steal your dog's food and dive-bomb your truck windows" (2013, 7). We bring birds to our waste sites, where they feed off all the stuff of our lives that we want to forget: "their success is due to our presence," as Gavan Watson (2013) puts it.

The gulls, *tulugaq*, and other species of birds who live on the trash heaps of human consumption may serve as a window into our production and consumption patterns and lifestyles, and how we understand ourselves in relation to other people, objects, and the environment. In a remark that points equally to inhuman and human alike, Greg Kennedy notes, "'trash' means a manner of physically relating to other beings ... We exist, for the most part, in a way that violently negates beings rather than takes care of them" (2007, xvi).

Indeed, news reports have documented Inuit Elders and other Inuit having to eat expired foods directly from local garbage dumps. This is perhaps unsurprising, as food prices in Nunavut are the highest in the country; all foods come highly packaged and must be flown in or shipped from the south. While many suggest retailers are benefiting from government food subsidies at the expense of people living in the community, this situation points to larger issues of Canadian Arctic development. As Madeleine Redfern, an Inuk woman and former mayor of Iqaluit, stated, "clearly people don't have enough money to be able to feed themselves" (in Rennie 2014a).

As the City of Iqaluit grows (the community has doubled in population over the last two decades), *tulugaq*'s relation to waste has come to represent a larger sustainability issue. A long-term Inuk resident of Iqaluit linked *tulugaq* to larger issues of community living in the Arctic:

> The amount of ravens there are in Iqaluit – it's disturbing ...
> The amount of bird droppings there are on the buildings all over Iqaluit is disturbing ... [My ancestors] wouldn't stay in one area. They would migrate with the animals so that they would sustain their own life. They wouldn't be in one area for very long because that food would be gone ... [Iqaluit is not sustainable] at the rate it's going. There's too many cars. There's too much garbage. There's too much [*sic*] people to sustain itself.[6]

For this resident, the congregation of *tulugaq* (and humans) *en masse* is itself disturbing. Perhaps, then, as Lathers puts it, within settler colonialism, "native shit [or in this case, *tulugaq* shit] reveals the failures of the new nationalism," exposing changing relationships with the inhuman and the tenuousness of community living (2006, 419). Importantly – and as these examples of Arctic trash animals make clear – waste itself readily informs how and where humans engage in multispecies relationships (ibid.). In Canada's Arctic, these inhuman/waste relations are developing in a context of competing stakeholder interests and rapidly changing communities. Like others living in the Arctic, *tulugaq* the Trickster is adapting to its settler colonial inheritance.

## QIMMIIQ

Inuit consider the killing of thousands of Inuit sled dogs (*qimmiit*) during the mid-twentieth century by southern Canadians (government workers, RCMP officers, and teachers) a "flash point" of colonial trauma (McHugh 2013, 42). These deaths occurred as part of a governance strategy aimed at changing Inuit relationships with the environment – a strategy that identified sled dogs as a kind of dangerous waste. The decision was enabled, at least in part, through an understanding of nature (sled dogs) as commodity – in this case, as a hunting tool easily replaced by new technologies (for a detailed discussion, see Tester 2010a; 2010b, 129–47).[7] The killings came out of this reconfiguring of sled dogs as trash animals, and profoundly influenced the shift for Inuit from a nomadic lifestyle to a sedentary, labour-based lifestyle.

For millennia, the bond between *qimmiit* and Inuit was integral to Inuit survival: when out on the land, *qimmiit* pulled sleds for hunting and for moving families, and weakened polar bears and muskox for hunting; when navigating sea ice, *qimmiit* detected potentially fatal soft patches on the ice; while out harpooning seal, *qimmiit* knew to keep quiet; and while Inuit families slept, *qimmiit* warded off predators (QIA 2013d, 9–15). During prolonged periods of starvation, Inuit ate *qimmiit* (the final step before eating leather clothing, tents, and dog sled lines) and used their pelts for clothing (ibid., 327). For Inuit hunters, survival required killing enough food to support both their families and their *qimmiit* teams; strong and well-fed *qimmiit* teams were a marker of manhood (McHugh 3013, 157). While fiercely loyal to their Inuit families, *qimmiit* were aggressive to other humans and often to each other – a characteristic essential to the success of *qimmiit* as hunters (Adamee Veevee, in QIA 2013e, 61). The training of *qimmiit*, who were considered neither wholly domesticated nor feral, was an ongoing task that required long-term *Qaujimajatuqangit* (Inuit knowledge) (ibid.).[8] Inuit boys were given *qimmiit* puppies to raise; training them socialized young Inuit as much as the dogs:

> in our customs there were a lot of regulations, though it seems typical that the Inuit don't have regulations. But in spite of that assumption, we did have a lot of regulations. For example, in raising [a] dog team, while they're still puppies we had to stretch

the legs, and rub their underarms, tickle them in order for them to get used to the harnesses, we did that during summer. While they're becoming adolescent dogs, we would have to take them for walks with their harnesses on ... We would make them run with their harnesses on, in order to keep them fit. If the *qimmiit* are not tamed that way they cannot be part of a dog team, they would not know how to run appropriately, they would be stubborn. (Sakiagaq 2009, 11)

Thus, raising *qimmiit* as part of a hunting team was a process of developing a mutual relationship between humans, animals, and the community. Indeed, the Inuktitut word *qimutsiit* defines the point at which Inuit and *qimmiit* hunting teams become "irreducible" (ibid.). Trained as *qimutsiit*, Inuit boys were considered men only when they were able to successfully support a full *qimmiit* team (Veevee, in QIA 2013b, 38; QIA 2013d). Perhaps as a result of this close and unique connection, *qimmiit* were the only animals other than humans to be given the names of the deceased. The Inuit naming practice (which persists today) ascribed the deceased's attributes to newborn Inuit or *qimmiit* (Stevenson 2012). Through this practice, Inuit and *qimmiit* shared complex kinship systems: *qimmiit* were aunts, uncles, cousins, siblings, grandparents, and former *qimmiit*. For some Inuit, *qimmiit* were "everything" (QIA 2013d, 15).

Before colonization, *qimmiit* were not recognized as animals in the Western sense of the term – as pets, companions, beasts of labour, commodities, property, or waste – though they were now considered as such by the southern Canadians and RCMP officers who governed these new Arctic communities. On 20 January 1949, under the premise of public health and safety, the government of the Northwest Territories (which had no Inuit representation) legally enacted the Ordinance Respecting Dogs (Croteau 2010, 133; QIA 2013b; RCMP 2006, 7; Tester 2010a, 134). The law prohibited *qimmiit* from running freely in communities. Any dogs caught roaming could be seized or destroyed at the discretion of RCMP officers (ibid., 135). As a result, Inuit living in communities were forced to tie up *qimmiit* (and in a specified manner, with metal chains) – a type of confinement that disrupted traditional rearing practices, and often proved impossible because chains and collars were not always available in community stores and were expensive (Croteau 2010, 11).[9] Sedentary living also prevented Inuit from following migratory animals, which made feeding *qimmiit* with country food practically impossible. When not tied up, hungry *qimmiit* could wander into community dumpsites and feed from waste, and these *qimmiit* were often much healthier than those that were chained (McHugh 2013, 167). A 1959 report from Arctic anthropologist Toshio Yatsushiro describes the difficult decisions Inuit were forced to make:

> The Eskimos understand, if they [*qimmiit*] are free they will be shot, but if they are tied they cannot get food, so maybe they will die anyhow. Eskimos bring food and water to the dogs when they have it, but often they don't have it. So when the dogs go free they eat garbage – when the RCMP saw it they shot them [*sic*] it is not good. (QIA 2013d, 43. Note that I do not endorse the term "Eskimo.")

The *qimmiit* killings began in the late 1950s after a period when the Canadian government required all Inuit children to be educated within a southern Canadian educational system, and parents settled in communities in order to remain close to their children, bringing hundreds of *qimmiit* with them (QIA 2013d, 30–31, 67; Croteau 2010, 138). Government and industry alike sought Inuit for low-paying labour but found that Inuit would leave these casual low-paying jobs in order to hunt and harvest on the land for extended periods of time. Killing *qimmiit* removed this "barrier" to industry and government profit. Once a hunter's *qimmiit* team was killed, hunting became impossible, and many Inuit were literally trapped in government communities (QIA 2013b, 17).

The conviction that the *qimmiit* killings were a government conspiracy, to at best assimilate Inuit into southern Canadian modes of living, or at worst eliminate Inuit entirely, is commonly expressed throughout Inuit territories today (ibid.). As Issacie Padlayat explains, "The governments tried to eliminate us by eliminating the dogs we depended on for survival but fortunately the Inuit are able to withstand hardships" (in Croteau 2010, 11). For many Inuit, whose lives until this point had been thoroughly integrated with the lives of dogs, the *qimmiit* killings opened up the very real fear that the police would also murder Inuit (in Stevenson 2012, 604–5).[10] Witnessing the killing of their kin relations became a source of anxiety and depression for many Inuit. But perhaps most importantly, these killings threatened what it meant to be Inuit: to raise

Figure 6.2 | *Qimmiit* and a raven on the Iqaluit dog lot moved to make way for the expanding Iqaluit airport.

*qimmiit*, to hunt, to experience oneself *as* the land. With RCMP officers enforcing laws predicated on settler colonial regimes to civilize and commodify nature, there was, as the Qikiqtani Inuit Association points out, "no need for a conspiracy" (QIA 2013e, 41).

In a particularly revealing letter dated 8 October 1958, Sergeant J. H. Wilson – an RCMP officer stationed in Quebec's Nunavik region – reveals some of the misunderstandings held by southern Canadian officials that contributed to the *qimmiit* killings. In the letter, Wilson disagrees with Inuit claims about no longer being able to hunt, writing that "this is in fact not correct as there has always been many more dogs here than are needed" (Croteau 2010, 13). In his view, dog teams were easily replaceable – simply go to the dump and grab any number of stray dogs. For Wilson, like other RCMP officers, sled dogs were strongly associated with waste: "most of the Eskimos cannot resist the temptation to let their dogs run loose with the hope that they will survive on the garbage from the RCAF [Royal Canadian Armed

Forces] Stn" (ibid., 13; note that I do not endorse the term "Eskimo"). For Wilson, caring for *qimmiit* properly was entangled with southern Canadian understandings of civility that required a display of ownership and control over animals that did not eat (RCMP) waste.

In 2006, the RCMP launched an internal enquiry to determine whether the RCMP had engaged in a federally mandated slaughter of *qimmiit* in the 1950s and 1960s. The self-investigation denied that any mass culling had occurred, and among its key findings was that "The Inuit sled dog is a large and aggressive animal that can pose a danger to public safety" (ibid., 14). Moreover, the report concluded that Inuit sled dogs had primarily been killed due to a combination of "epidemics and socio-economic factors," the latter including the "social benefits to which the Inuit people had access for the first time, including government education, healthcare, government housing, and government family allowances within settlements ... [and] the introduction of the snowmobile" (ibid., 15). The overarching point – that the killing of *qimmiit* by settlers ensured the year-round availability of a cheap Inuit workforce, and more generally aided the settler colonial project – is obscured by the RCMP report's focus on education, and health care benefits. This focus also skirts the serious criticisms of the availability and quality of the education, health care, and other southern systems that southern Canada introduced to the Arctic, and through which the Canadian government still governs Inuit.

By contrast, the in-depth investigation into the *qimmiit* killings by the Qikiqtani Truth Commission (QTC), which involved interviewing hundreds of Inuit and *Qallunaat* Nunavut residents, concluded that the *qimmiit* killings occurred "because Qallunaat were scared of dogs" (QIA 2015, 39).[11] The fear of *qimmiit* was a loaded one, entrenched in southern Canadian understandings of cleanliness, civilization, and safety. As with the fear of diseases spread by landfill gulls in the south, the killing of *qimmiit* was, according to the QTC, "more about what they [*qimmiit*] might do in the future" (QIA 2013d, 53). The colonial rhetoric of safety and security, which scholars have argued is commonly used to justify both neoliberal governance regimes and state violence, was, and continues to be, employed by Canadian government officials to legitimize expansion throughout Canada's Arctic (Agamben 2002).[12] As with efforts to exterminate trash animals in Canada's south, managing perceived forms of disorder in the Arctic remains a central tenet of settler colonialism and management practices.

## INHUMAN SETTLER COLONIALISM

Although *qimmiit* are now the official animal of Nunavut, only about 300 remain in Canada – a far cry from the some 20,000 that existed prior to settler colonization (McHugh 2013, 158). In Iqaluit today, *qimmiit* are protected by law, albeit primarily for the purposes of conservation ("the Canadian Inuit Dog is the last [I]ndigenous dog to North America") and commodification ("Canadian Inuit Dog and Dog Teams ... are an integral part of the City of Iqaluit's unique character and its economic and tourism development") (Corporation of the Town of Iqaluit 2001, 1). Within city limits, *qimmiit* are restricted to "Designated Dog Team Areas," where all *qimmiit* must be tied up, tagged, and registered with the local authorities (ibid., 6). In stark contrast to the *tulugaq* at the dump, Iqaluit's *qimmiit* are disciplined, tidy, and orderly (see also De Vos 2013, 286).[13] These materially reworked relationships with the inhuman make up Iqaluit's colonial inheritance. *Qimmiit* have been, as Belcourt puts it, re-configured as "neoliberal subjects" (2015, 6).

Yet despite their utility within neoliberal capitalist modes of production, sled dogs are once again being threatened (Patriquin 2015). As part of the Iqaluit airport's $300 million expansion, which opened in May 2017, one of the community's three remaining *qimmiit* lots was placed off-limits and replaced by an asphalt plant. The expanded airport is a response to increased southern Canadian and international industrial and military activity in the region, and Iqaluit residents are now concerned about where to put their dog teams (Varga 2015). As a local dog musher explains, "If it becomes too much harder to have dog teams, people are just going to give up, and it'll die off" (Patriquin 2015). The dog lot in question is located along a 330-metre stretch of road on a former airport landing strip. Despite being located centrally between at least five dumpsites (including the old and current community dumpsite, as well as three military dumps), the lot is considered by many to be the safest place to keep their *qimmiit*.

One of the other *qimmiit* lots is on the banks of Iqaluit's airport creek, a site known for its high concentrations of carcinogenic chlorinated paraffins, likely leaching into the water from an abandoned up-creek military waste site (Dick, Gallagher, and Tomy 2010). A *Qallunaat* dog owner explains why she recently moved her dogs away from airport creek:

I just started realizing that I've had a few dogs die before the age of ten. And one of them, they had – I mean, I don't know exactly what was wrong with them, but she had her – her liver was totally not the right colour, not the right texture. All of that stuff. So – and then I had another dog die of prostate cancer, which is not really that normal for an unneutered male. And then I had another dog die – also around that same age – of sort of unknown causes but [it presented symptoms that were] sort of similar to the one that I did know had a screwed-up liver ... And so all the dogs that I had that didn't grow up on that creek lived to be like 14, 15. And suddenly [there is] this whole generation of dogs that I had [that] seemed to be dying younger than I would have thought. And, like, you can't really draw that link, but it makes you think.[14]

Waste, both symbolically and materially, has become part of Nunavut's settler colonial present. Like other social and environmental issues developed out of settler colonial governance, waste histories haunt present and future generations. Despite territorial government efforts to incorporate traditional Inuit knowledge into territorial and municipal government decision-making processes, both the act of wasting and the management of waste are governed through southern Canadian structures, processes, and practices. Waste is, in many ways, itself colonial, as this book argues. Nunavut communities are now the largest producers of waste in Canada's territories (Van Gulck 2012, 24).

This examination of Nunavut's trash animals demonstrates how ravens and dogs, like Inuit themselves, have been incorporated into, and managed by, historically, culturally, and materially constituted cosmologies. Within our current neoliberal capitalist governance structure, waste is inherited as both a material and symbolic form of disorder, unruliness, and disgust. Referring to David Gilmartin's analysis of British irrigation engineers in the Indus Basin, Baviskar notes:

Controlling waste was, in differing ways, crucial to both an agenda of increasing "scientific" control over the environment, and to the state's political manipulation of [I]ndigenous communities. Understanding the place of waste in colonial discourse is thus a way of understanding some of the most basic contradictions underlying this resource regime. (2003, 5,053)

Disorderly relationships with Inuit, animals, and landscape have been variously embraced, compromised, co-opted, commodified, destroyed, rebuilt, and abandoned. In Nunavut, as elsewhere, we are tasked with the challenge of inheriting an increasingly messy, uncertain, and colonized "lifeworld" – one shaped by climate change, toxic and indestructible wastes, and indeterminate human and environmental impacts. The effort to name our current epoch the Anthropocene is (as the next section will discuss), among other things, a bid to formally acknowledge the association between neoliberal capitalist modes of governance, Western scientific modes of knowing, and our ecological and settler colonial inheritance. In this context, it is important to critically examine, question, and challenge the political, historical, and cultural structures through which our inhuman relations are variously practiced and embedded. Learning how to inherit, then, becomes a matter of reimagining and thus materially reworking relationships with the inhuman in ways that accept rather than dominate other lives and livelihoods – and to go further, that challenge waste strategies that promote neoliberal capitalist governance and settler colonialism at the cost of community health and environmental well-being. Trash animals continue to remind us – however briefly – of that which we can never truly abandon or forget. Though we may choose to ignore these animals or even legislate their disappearance, they will inevitably show up in our imagined sanitary lifeworlds.

# SECTION THREE

# Waste at the End of the World

CHAPTER SEVEN

# A Good Soup

## THE FLIP-SIDE TO EXTINCTION

Section 1 (Waste in Southern Canada) and section 2 (Canada's Arctic Wastes) provided theoretical and empirical evidence concerning the state of waste management (WM) in southern and Arctic regions of Canada. In this next section, Waste at the End of the World,[1] I move the discussion to a more theoretical register that engages with both the indeterminacy of waste, and waste as a physical signature of the Anthropocene. This engages an important scalar effect from local actions to global effects. Chapters 7 and 8 demonstrate that waste is not only a *local* concern that cannot be meaningfully understood isolated from social justice concerns, but a *planetary* concern. That is, whereas the chapters in sections 1 and 2 focused on the political, cultural, economic, and social forces that produce our current waste crisis, chapters 7 and 8 demonstrate that waste is also *material*, and this materiality conditions how waste itself engages with the planet. I explore this argument by showing how, despite human efforts to control waste (by governing in particular ways), it nevertheless proves to be rather stubbornly indeterminate (as civil engineering and biological studies of leachate, toxic compounds, and so on detail). For many years, Nigel Clark and I have shared an interest in biogeological planetary forces; I spent years studying bacteria as progenitors of all other life on Earth, and Nigel has focused on inhuman forces such as fire. Chapter 7 is the outcome of one of our many discussions about the exuberance of inhuman life. With an appreciation of the materiality of waste in chapters 7 and 8, I return in chapter 9 to demonstrate how local residents in Iqaluit were able to mobilize the indeterminate materiality of the West 40 dump and fire to convincingly argue against the Canadian federal government's declarations about the risks of the dump fire to human health and the environment. In this way, chapter 9 assembles all of this book's major themes – section 1's focus on waste governance in southern regions, section 2's focus on the impact of ongoing settler colonialism on waste governance in the Arctic, and section 3's focus on the material limits of the human governance of waste.

An observation that is as awe-inspiring as is it obvious: every living creature on this planet belongs to an unbroken chain of bodies that wends its way through time, from one birthing or fission to another, all the way back to the emergence of life itself from non-living matter. Each bacterium, each angiosperm, each vertebrate is part of a continuous thread which neither continental nor genetic drift, volcanic winter nor snowball glaciation, meteor impact nor paleo-atmospheric crisis has been able to sever. However much we celebrate non-filial transfers and couplings, the lateral interchanges of networks and assemblages, or the indefatigable flux of life, any snuffing out of a genus or species of living beings, any severance of this 3.5 billion or so years of living is worth taking seriously. Extinction matters. Some consolation may be found in the capacity of inanimate matter to self-organize into more complex arrangements. However, it is worth dwelling on the fact that the mass of once-living organisms that have died and returned to base matter has been estimated to be somewhere between one thousand and ten thousand times the mass of the Earth itself (see M. Davis 1996), while as far as scientists have been able to ascertain, matter has only organized itself into life once on our planet. There is, then, something of an asymmetry between the transition from base matter to life and the traffic that passes in the inverse direction, from living back to base matter.

Even the practice of anticipating, documenting, and mourning the passing of a species, the melancholic publicity so pivotal to contemporary environmentalism, falls short of embracing the profundity of extinction's loss. For, as Kathryn Yusoff (2012) reminds us, a great many of the living lineages that are now likely to be disappearing have never been seen by us in the first place: their deaths take place unattended and unannounced, beyond the capabilities of the technologies we use to detect them. Too small, too obscure, too reticent to have graced our archives, these beings blink out of existence without ever making their presence felt. If there might be such a thing as a dark ecology, a green so deep as to emit no glimmer of light, perhaps it lies here, in the contemplation of the finality of a withdrawal from the ranks of those living beings we never even knew existed.

But what if there is a flip side to the anonymous eclipse of so many species or strains? What if, without trying, without knowing, without even the possibility of our finding out, we humans were *increasing* the sum total of biological diversity on Earth? The Anthropocene, it turns out, may be as much about species proliferation as it is about species extinction. Moreover, as much as humans have become a geologic force, we do not understand, nor are we capable of predicting, let alone controlling, our impacts on the planet (see Hird 2009).

Take, for instance, the example of watermelon snow. When Martyn Tranter, a biogeochemist conducting research in Greenland's Kangerlussuaq region, forgot his glacier goggles and resorted to wearing tinted cycling glasses, he was surprised to see a kaleidoscope of mauve, green, red, and brown enlivening the ice sheets on which he was working (Witze 2016). While the phenomenon of chromatically tinged snow has been observed for thousands of years, it has only recently been understood that proliferating microorganisms produce these colour effects as they awake from hibernation during relatively warmer temperatures. As the global climate warms and polar ice sheets melt, conditions become increasingly hospitable for algal and bacterial species, as Tranter recognized. But at the same time, microbial ice colouration increases sunlight absorption, which in turn accelerates melting (Witze 2016).

Just as scientists are discovering the positive feedback potential of the ice-dwelling algal blooms on global climate change, they are also realizing that this gigantic melting process is liberating unknown numbers and kinds of microbes that have been effectively cryogenized in deep ice reservoirs in both the Arctic and Antarctic (Katz 2012). Microorganisms frozen in the ice during glacial cycles that go back at least as far as the mid-Pleistocene epoch are now thawing out and re-entering the biosphere after a break of up to a million years. Already, microbes taken from layers of ice laid down over 400,000 years ago have been successfully grown in laboratories (ibid.).

While scientists believe these cold-loving organisms pose little immediate threat to the health of larger, warm-blooded creatures, they are concerned about their impact on marine ecosystems and on the Earth system more generally. One threat scientists have been considering is that a flood of organic nutrients released from the breakup of ice sheets could trigger blooms of contemporary bacteria that use up oxygen in the water, contributing to ocean dead zones. Another risk arises out of the possible decomposition of so much organic matter – the biomass of microbes in and beneath polar ice being estimated to be over a thousand times that of the Earth's human population – which would add a massive surcharge of carbon dioxide and methane to the already escalating greenhouse gas composition of the planet's atmosphere (ibid.).

Even less predictable are the consequences of ancient microbes joining the current web of life. Though they have been shaped by long-gone terrestrial conditions, revivified Pleistocene microbes will begin evolving afresh, and will once again exchange DNA with their fellow microbes – both old and new. As evolutionary biologist Scott Rogers explains: "What we think is happening is that things are melting out all the time and you're getting mixing of these old and new genotypes" (in Katz 2012, unpag). And nobody, it seems, has any idea of the ultimate consequences of this exchange between life forms once believed to be extinct with those that have evolved to live on the Earth as we know it.

Microbes are capable of reproducing and differentiating very quickly, are highly adaptable and capable, and are adept at creating new life forms. As evolutionary biologist Lynn Margulis (1998) liked to remind us, life on Earth was, is, and most likely will be as long as it endures, thoroughly dominated by microbial communities. For the first 2.5 billion years of terrestrial life, single- and multi-celled organisms from the Archaea and Bacteria domains were the only life forms around: a microscopic throng whose proliferation, promiscuous exchanges, and evolutionary radiation collectively generated our solar system's only biosphere. Bacteria invented all the basic metabolic processes, including photosynthesis and chemical conversion, that every other life form remains utterly dependent on (Hird 2009). And it is this peerless proficiency at metabolizing available matter-energy, everything from solar radiation to organic matter, and metallic ores to acidic sulphates, that makes bacteria so important when it comes to disposing of our own stockpiles of surplus and unwanted matter.

This brings us to the ecological predicament of most of the planet's human communities. The waste products of human productive and consumptive activity need to go someplace, and the exponential rise in productivity over the brief span of industrial modernity means that those places must be evermore capacious. Our counterpart to the wholly organic jetsam of our ancestors is a concoction of inorganic and organic detritus, and our answer to their middens is a trans-global multiplication of landfill sites. Open dumping and landfilling is, globally, our preferred means of waste disposal. Eventually, whatever we stash underground comes into contact with the bacterial life that dwells in the soil – or rather, given their populace of some 40 million per gram, we might say they *are* the soil. Bacteria do what they have been doing since the Eoarchean era: they figure out ways of metabolizing whatever matter-energy they encounter. Thus, each landfill is, in its own way, a unique bundle of materials, at once an ancient and a novel challenge to bacterial communities. As the next chapter will detail, the landfills of contemporary industrial societies include various amounts and kinds of seven million or so known chemicals (and the thousand new chemicals that enter into use each year), along with a full spectrum of organic matter, which includes the 14,000 food additives and contaminants found in our food scraps. The liquid material or leachate into which organic landfill dissolves frequently consists of a heterogeneous mix of heavy metals, endocrine-disrupting chemicals, phthalates, herbicides, pesticides, and various gases, including methane, carbon dioxide, carbon monoxide, hydrogen, oxygen, nitrogen, and hydrogen sulphide.

We simply have little idea as to what bacteria ultimately make of these and other ingredients, and what in the process they make of themselves. This is not a question of more control, of more techno-science and more governance. How bacteria will adapt, and into what new life forms bacteria will differentiate is inherently indeterminate. The creation of dumps and landfills, and waste more generally, is not the only way in which human activity incites the proliferation and transformation of bacterial life. However, the unfathomably rich and complex feedstock that we are pumping underground has a special significance in the magnification of what we do not, and cannot, know, and its unintended consequences comprise one of the deepest and darkest ecologies of the current material-historical juncture.

A quite feasible, but utterly unconfirmable, consequence of human subterranean waste disposal is a stimulation of bacterial proliferation that is likely to involve adaptation and diversification. Given the vast populations and the huge variety of bacteria, such augmented diversity might even exceed the accelerating extinctions about which we are variously ignorant and hyper-informed. However, this is not a numbers game, at least not one in which the divisions, subtractions, or multiplications belong to us, nor is it one which is solely or even primarily about interconnectivity, networking, and entanglement. The biggest ontological provocation of the conjunction of human waste and bacteria is the fact of our total dependence on life forms whose life-worlds and trajectories are likely to remain overwhelmingly unknown to us. If this provides a cautionary note about our own increasingly intensified disturbances of the Earth's constitutive strata, and the waste that this produces, perhaps its more profound provocation is about the force of the stratifications and de-stratifications that the planet itself undertakes.

## DEEP SHIT

Social analyses of the accelerating movements across the surface of our planet, encounters, and complex relationships that go by the name of globalization have frequently drawn on Bruno Latour's (2005) notion of multi-actor networking, along with Gilles Deleuze and Félix Guattari's (1980) take on territorial dynamics. Recent social scientific engagement with human waste covers this general terrain, tracking the circulation of assorted human refuse through increasingly globalized networks (for example, Canadian waste ending up in the Philippines) and showing how this occasions more or less unpredictable transformations of both the waste products themselves and the networks and environments through which they flow.

This is in keeping with the rise of various theories of relational materiality (discussed further in the next chapter). The priority of this approach is to bring to light spatial orderings and transformations involving a range of human and inhuman participants. Yet however stretched out, reticulated, or enfolded we imagine the spaces in question to be, there has been a pronounced reluctance to prise beneath the uppermost stratum of human activity over the last few decades of critical social thought. While a multitude of objects have been followed across the globe, this pursuit has rarely extended far into regions that are devoid of human presence or indifferent to human entreaty.

In the case of the subsurface depositing of waste, there is a need to delve beneath the relations that compose or decompose territories, and to burrow into the underpinnings of the organization and patterns found on the surface of the Earth. Deleuze and Guattari's (1980) notion of strata, that of relatively consistent layers or belts of substances that compose the body of the Earth, has not quite attracted the same enthusiasm as their concepts of de-territorialization and re-territorialization. Where Deleuzo-Guattarian strata have been featured in social and philosophical thought, it is usually with regard to their capacity to cut across each other or to be traversed and mixed. Deleuze and Guattari's own foregrounding of de-stratification offers a necessary corrective to timeworn tendencies toward foundational determinisms. But it is worth recalling that, although they rejected any simple unidirectional or teleological relationship between one stratum and the next, they broadly accepted that an earlier stratum created the conditions of possibility for strata that followed, proposing "a coded system of stratification" made up of "hierarchies of order between groupings; and, holding it all together in depth, a succession of framing forms, each of which informs a substance and in turn serves as a substance for another form" (Deleuze and Guattari 1980, 335).

Biological life, Deleuze and Guattari noted, forms a stratum of its own, but at the same time has a particular propensity for de-stratifying, for traversing and disturbing the geological layers that are its underpinning (1980, 336). Human life, too, the "anthropomorphic stratum," participates actively in the crosscutting and back-blending of other strata. As Deleuze and Guattari famously put it: "it is possible to reverse the order with cultural or technical phenomena providing a fertile soil, a *good soup*, for the development of insects, bacteria, germs, or even particles" (1980, 69, emphasis added).

One of the most public manifestations of this "possibility" in recent years was the 2010 Deepwater Horizon oil spill. Microbes do not play a significant or even supporting role in the 2016 movie *Deepwater Horizon*, which understandably focuses on the human tragedy and drama surrounding the 2010 explosion on the eponymous drilling rig. Yet, as seepage from the uncapped wellhead grew into the world's largest accidental oil spill, and concern over impacts in the Gulf of Mexico escalated into one of the worst ever environmental crises, microorganisms came to play a pivotal part in the narrative. In this event, massive volumes of hydrocarbons extracted from deep beneath the seabed were accidently released into the different vertical layers of the ocean, the oil plume entering various webs of marine life inhabiting these layers. For scientific observers, what made the Deepwater Horizon spill unique was not only its scale – an estimated 4.9 million barrels and a slick covering 112,000 square kilometres of the ocean's surface – it was also the fact that a massive plume of hydrocarbons reached a depth of around 1,100 metres (Kimes et al. 2014; Beyer et al. 2016).

With the oil spill's widespread impacts on marine life and coastal ecosystems, including its devastating implications for human livelihoods around the Gulf, the accident and the waste it produced soon escalated into a political controversy in which questions of culpability and compensation were fiercely debated. These debates were complicated by the locality of the oil discharge – in particular by its position in the deep sea: or what Phillip Steinberg and Kimberley Peters depict as "the ocean itself ... its three-dimensional and turbulent materiality" (2015, 247). These debates soon came to include the role of biological life – not simply marine ecosystems in general, but the specificities of living communities inhabiting different depths and layers of the ocean. To what degree would marine microorganisms be able to consume and degrade the submarine hydrocarbon plume? With a speed and efficacy that would likely have surprised even Louis Pasteur, bacteria that few people even knew existed became major actors in the political spaces that took shape around the spill. In this case as in others, the techno-scientific promise of deploying bacteria to metabolize and somehow render inert our toxic waste is becoming ever more popular.

Although offshore oil drilling had been taking place in the Gulf for over fifty years, and despite the fact that bacteria have been consuming naturally occurring hydrocarbon seepage in the marine environment for hundreds of millions of years, prior to the disaster, "relatively little was known about northern Gulf of Mexico bacterioplankton" – especially with regard to their diversity (King et al. 2015, 379). Early scientific evidence suggested that blooming populations of microorganisms were making short work of consuming the hydrocarbon plume (Hazen et al. 2010). "The microbes did a spectacular job of eating a lot of the natural gas," concluded biogeochemist Chris Reddy, adding "The rate and capacity is a mind-boggling testament to microbes" (cited in Biello 2015, unpag). In a widely reported public lecture at the University of Southern Mississippi entitled, "Can Mother Nature Take a Punch? Microbes and the BP Oil Spill in the Gulf of Mexico," microbiologist Terry Hazen answered largely in the affirmative, recounting how bacteria that were pre-adapted to the natural presence of hydrocarbons in their ecosystem had thronged to the plume like "oil-seeking missiles" (cited in Kirgin 2011, unpag). Unsurprisingly, this was a narrative that appealed to powerful vested interests in the oil economy. Six years after the event, Geoff Morrell, senior vice president of BP, the company that had operated Deepwater Horizon, picked up on the microbial remediation narrative, insisting: "There is nothing to suggest other than that the Gulf is a resilient body of water that has bounced back strongly" (cited in Elliott 2015, unpag).

But the superficial good news story that the oil industry, politicians, and some of the media embraced turned out to be, well, a story. Indeed, the success of the microbial cleanup has been strongly contested. Even after the publication of over five hundred scientific research papers on the environmental impacts of the spill, there remains much uncertainty about the response of microorganisms to the massive hydrocarbon influx and its ecological consequences (Beyer et al. 2016; King et al. 2015). Earlier research suggested that the enhanced oxygen consumption of the bacteria that consumed the spilled oil could be severely affecting photosynthesizing marine bacteria (Widger et al. 2011). As time went on, one of the major unanswered questions was what became of contaminants as they moved through, and up, the marine food web – as hydrocarbon-consuming bacteria, archaea, and microfungi were in turn consumed by other organisms (Beyer et al. 2016). But this kind of profound and lingering uncertainty is difficult to get on – and keep on – the political agenda, and the Deepwater Horizon oil spill soon had to compete with other environmental disasters for our attention, as was detailed in chapter 2 (Downs 1972).

What might this tell us about the political geographies of emergent life in the contemporary world? Bacteria that willingly consume oil pollution and potentially pathogenic microorganisms have at least one important thing in common: they are both bound up with human actors in ways that can be, with a little effort, brought to light. Their worlds and our worlds are mutually implicated, or in the language of relational ontologies, *co-enacted*, as Clark puts it (2011, 30–4). So, too, it has been argued, are we humans entangled with the ocean itself, a body that Steinberg and Peters describe as "a volume of vibrant matter that is enlivened and made forceful through its *relation* with human life" (2015, 256). But a big part of the ontological trouble with oil-consuming bacteria as they or their residues negotiate complex marine food webs is that any such relation to humans withdraws deep into the distance.

From the perspective of marine organisms and from the vantage point of the many human actors – who will unlikely ever fully access the ecosystem dynamics in question – the relationships that count have effectively ceased to include humans (N. Clark 2011). In other words, the ultimate effects of the 2011 Gulf oil spill on marine life, both micro- and macroscopic, remains fundamentally inconclusive. In many senses, the practice of putting human refuse underground might be viewed as a kind of terrestrial Deepwater Horizon on a massive scale, both spatially and through time. Whereas the relatively narrow spectrum of the marine bacterial phyla that benefitted from the Gulf spill were fuelled by a fairly uniform feedstock of petroleum hydrocarbons, waste matter intentionally deposited in landfills is generally characterized by a heterogeneity of ingredients unique to each individual landfill.

Depending on its location and the period in which it was operative, any given landfill might contain chemicals whose health effects and environmental consequences are well-known – chemicals that have since been prohibited or are banned elsewhere – and those whose long-term consequences are scarcely known at all. The various cells that compose a modern landfill may themselves have a very different composition, an inconsistency that is in turn greatly intensified when we consider the wildly uneven array of disposal contents, practices, and regulations across the globe.

At least in some highly regulated landfill sites in the West, sophisticated techniques of lining landfills will extend the period of containment. However, as the previous chapters have indicated, containment is ultimately imprescriptible, given a long enough timeline, which raises the issue of how microorganisms in the surrounding soil and water will respond to the leachate that sooner or later will seep from landfills. We do not, and cannot, ultimately know which bacterial taxa are present, which populations will be deleteriously impacted by the specific mix of chemicals and organic materials they are exposed to, and which will adapt and proliferate under novel conditions. There is no tally sheet keeping track of microbial diversity that has been extinguished, alongside newly created microbial diversification and stratification. This is Yusoff's (2012) invisible, nonpresentable extinction, entangled indistinguishably with an equally nonpresentable becoming.

## BACTERIAL DE-STRATIFICATION

Whether conceived in terms of Deleuzo-Guattarian *agencement*, Latourian networking, or more classical ecological systems, the interconnectivity among bacteria, and between bacteria and other entities, would seem to be an exemplar of inhuman object relations. Bacteria have largely bypassed the territorialization that is referred to as speciation in regard to so-called more complex organisms. Their reproductive habits are as wildly variable as their capacity to form symbiotic couplings with other organisms of varying scale and complexity (Hird 2009). Alongside and in collaboration with such geological forces as sedimentation, volcanicity, and the ocean-bed upwelling of magma, bacteria are also inveterate composers of geological strata. The progenitors of the Earth's biosphere, bacteria shaped the anaerobic and later aerobic atmosphere and play a significant role in rock weathering and the production of metallic ores, as well as provide the basic components of all other life (ibid.). Without photosynthesizing bacteria, there would be no reserves of fossilized biomass to fuel human industrial production (ibid.). Our oil industry is entirely based on the extraction of the dead mass of plankton that age and bacteria have metabolized into oil. Without the bacterial building blocks of multicellular life, there would be no humans to extract and set to work this subterranean reservoir of energy.

However much the materials that humans assemble and dispose of traverse the surface of the Earth, it is our accelerating capacity for re-stratification that the waste problem throws into relief. Like upside-down department stores or inverted hypermarkets, landfills thrust their discarded cargo into the planet's geologic layers. Subsurface flows of water extend this interjection, conveying leachate through the pores and seams of the Earth's crust, drawing it deeper into the stratigraphic column, effectively mixing the residue of our contemporary waste with that which existed long before us. But this is only half the story. The reinsertion of heterogeneous and composite materials into the Earth is a haphazard echo of an earlier set of de-stratifications: the prising of useful energy and raw materials out of geological formations. In the era of industrial production, most of what becomes landfill has in one way or another been forged from subterranean ingredients. Whether it is metallic ores tunnelled from the ground, plastics fashioned from fossil hydrocarbons, or dramatically increased food volumes made possible by petrochemicals, the better part of what goes (back) down has first been brought up.

## RE-STRATIFICATION

The reconceptualization of cities as information/digital infrastructures is made possible through a material infrastructure – cables, pipes, and materials involving metals and non-renewable fossil fuels – that forms connections within and between cities, countries, and nations. As Keller Easterling points out, "infrastructure is now the overt point of contact and access between us all – the rules governing the space of everyday life" (2014, 11). Developed as private international enterprises rather than public works, these heavy industries nevertheless "engaged in constructing the terms of civilian life – laying the cable, providing the rolling stock, building the canal or dam ... They often shaped legislation and determined the values that were worth defending militarily" (ibid., 152).

This, in turn, becomes a vast waste infrastructure when technologies break down and are no longer repaired or are updated or replaced. The discards of urban infrastructures – urks – join the global landfill and nuclear repository infrastructure meant to contain and disappear humanity's unwanted and abandoned objects in perpetuity – out of sight and out of mind. Since so much of the materials that make up urban infrastructures are already buried underground, urks are all the more easily forgotten. If the ghost is, as Jacques Derrida (1994) argued, the permanent return of the absent, then we might ask whether the resurrection of urks – and this is exactly what is on the industrial agenda – testifies to "a living past or a living future" (Gordillo 2014, 246). Relative to the sustained attention devoted to the sociotechnical aspects of infrastructure, the effects of its waste, emissions, land use, and contamination are rarely considered (Monstadt 2009), as the latter are concerned with destruction, disassembly, disconnection – what Gastón Gordillo (2014) observes is a hierarchy of waste such that what cannot be resurrected is an inferior form of matter (see also Graham and Thrift 2007).

We have put into use almost all of the metals in the periodic table (UNEP 2010), and industries are assessing the profitability of extracting nickel, copper, iron/steel, aluminum, and other metals from urban infrastructure waste in a distinct form of urban mining, making use of what some industries refer to as "hibernating stocks," since the amount of specific metals such as iron and copper in the built environment meets or "exceeds the amount in known geological ores" (Johansson 2013, 1; Bergbäck and Lohm 1997; Spatari et al. 2005). Pipes and cables constitute about a fourth of the weight of a city's infrastructure and in some cases contain as much or more sought-after metal as operating mines (Wallsten 2013). Resurrecting urks is a complex process, as cables and pipes laid down on top of each other over time through successive urban-space planning stages has created a vast, intricate, and difficult-to-access material system. Deleted files, abandoned maps, and the generally haphazard record keeping of successive local governments and competing industries ensure that the recovery of desirable metals is both tricky and an uncertain venture (Wallsten 2014). Often the abandoned infrastructure is only discovered in the process of repair work to an existing and still reasonably functional infrastructure. This said, as contemporary titanic mining efforts (i.e., initial extraction) extract fewer and fewer sought-after minerals and metals, the increasing interest in urk mining highlights capitalism's long vertical as well as horizontal reach. Buried under layers of infrastructure that may well be destroyed in the process of extraction, "through a decisive sleight of hand, destruction is redefined as innovative, positive, desirable" (Gordillo 2014, 80).

Indeed, urks provoke ontological questions about the very conceptualization of waste as such: we are back to the profound importance of defining waste. If urks were to be defined as waste, then they would need to be removed and landfilled, constituting a near-impossible remediation project that would jeopardize the successive layers of infrastructure built on top of the abandoned urks. As such, urks tend to be defined as other entities – contaminated soil or resource, for instance – that skirts the pressing need to remediate, since other contaminated areas (such as abandoned oil refineries or chemical plants) are typically much more likely to leak and spread contamination (Wallsten 2014). Nevertheless, this definition of convenience must remain tentative, as certain urks may contain materials – PCBs, CFCs, lead – that may not stay put, may leak, and may require removal and remediation (ibid.). Moreover, there is always the danger of precipitating such a leak when urks are mined. And, of course, much of the material that will inevitably be resurrected with the sought-after metals and minerals will be redefined as waste and reburied in landfills.

As we know, despite well-publicized efforts to recycle, the majority of products collected by WM industries globally (which have assumed control of large parts of the recycling industry) end up in landfills (Kim and Owens 2010; UN-Habitat 2010). And as Johansson notes, landfills are "bursting with metals: globally over billions of tonnes of iron and millions of tonnes of copper, and other crucial resources" (2013, 33). As the second section of this book examined, there are numerous challenges: municipal solid landfills and even exclusively industrial landfills tend to contain a heterogeneous mix of diverse materials in a series of vertical cells built on top of each other, and gleaning the contents of each cell is difficult without detailed record keeping (which is uncommon). A number of WM and mining industries are conducting exploratory studies of the viability of this "post-mining" through the use of vertical bore holes that cut through vertically arranged cells. Landfill mining is precarious work, since it involves disturbing sedimented landfilled material, cutting through liner systems designed to hold leachate in place, and mixing materials, an unknown quantity of which may be toxic. It means reintroducing oxygen into the landfill system, which changes the aerobic and anaerobic bacterial metabolism of the landfill. We know little of the bacterial stimulation and proliferation that occurs in landfills, which is likely to involve bacterial adaptation and diversification. Thus, insofar as landfilling is a geologic re-layering of materials that have already been extracted, landfill mining constitutes a re-extraction, a further de-stratification and mixing of the "productions of deep, planetary time" that "defy our own [temporal and geologic] sensorium" (ibid., 49–50).

And somewhere in between burial and resurrection lies hydraulic fracturing. Fracking stands to significantly increase the already 57,000 million tons of material we are annually shifting from beneath the earth to its surface (I. Douglas and Lawson 2000). As we further disturb geologic strata, we are significantly adding to the biological soup that traverses and variously forms and responds to the geologic. Not only is fracking literally unearthing billions of several kinds of subterranean extremophile bacteria (able to thrive in intense conditions such as extreme heat that are catastrophic to animals), but in order to dislodge the bacterial colonies that are clogging fracking wells, companies are injecting millions of litres of biocides into the earth. The United States Environmental Protection Agency (EPA) has thus far approved over twenty-eight different biocides for fracking applications (Kahrilas et al. 2015; Sager 2014). These biocides may seep between rock beds, through surface soil, and into the water table. Not only do we not understand the effects of this seepage and contamination on human health, flora, and fauna, we have little idea of the consequences of the interactions between these extremophiles and the bacterial ecosystems in the soil, lakes, and oceans on which we depend for survival. Such resurrection, as Dr Frankenstein learned at cost, may well revive, or indeed create, more than we intend.

Deleuze and Guattari's (1980) cautioning about the destructive consequences of a too rapid de-stratification appears to be an understatement when we consider the current velocity of both extraction and reinterment, and then re-extraction. The bauxite that is the basis of aluminum derives from volcanic ash deposited tens of millions of years ago; the fossilized hydrocarbons from plastics or fertilizers has been hundreds of millions of years in the making; while an important source of commercial iron consists of banded rock laid down in the Precambrian era, over two billion years ago. Given that the functional use of a foil food wrap or drink bottle may last no more than seconds before it is discarded and enters the long *durée* of subsurface disposal, we are positioning ourselves in the midst of temporal disjunctions that defy our own ability to sense and perceive, let alone understand or control. In this sense, the current scale of the extraction of minerals, deep-sea oil, and unconventional hydrocarbons, together with the exponential rate of landfilling, might be seen as a kind of massive conveyor belt of de-stratification. This is traversing and mixing what has been produced by deep, planetary time, for which the largely spatial dynamics conveyed by the term "globalization" cannot begin to do justice. Landfills are supposed to be about re-stratification after extraction. Our unwanted materials are supposed to stay put. But we know that via leachate, as well as via a host of containment failures due to fires, landslides, explosions, and so on, the waste in landfills doesn't stay put. And now we are intentionally adding dump, landfill, and urk mining and fracking to the mix.

## ASYMMETRY

We may dig deeper and do our hefting in more spectacular quantities, but next to the traversal artistry of bacteria, humans are crude and clumsy de-stratifiers. To migrate across strata while extracting and reprocessing available elements is the forte of microbial life. Those miniscule metabolic engines whose negotiation between the most diverse formations of the Earth have been constantly fine-tuned and retuned over billions of years. Whether the accelerated de-stratification wrought by our own species over the geological eye-blink of the industrial age ultimately serves to augment or subtract from the largely microbial biological diversity of the Earth may not be the key point. As Isabelle Stengers puts it, in a voice that could be conversing with Margulis, "Of the Earth ... we can presuppose a single thing: it doesn't care about the questions we ask about it" (2000b, 145). And neither, most of the time, do bacteria, the planet's organic prime movers (Hird 2010a, 2010b).

And yet, what bacteria do with the substances to which we expose them, or what this exposure does to bacterial populations, may have profound consequences for humans and other organisms. In an empirical sense, we lack access to the vast majority of bacterial losses, gains, and transformations: dynamics that are obscured by the scalar mismatch of bacteria and ourselves – by the immensity of their numbers, strangeness of their forms, and the difficulty of accessing many of the environments in which they thrive. In an ontological sense, what it is to be a bacterium, or more appositely, a vast meshwork of interacting bacteria (that some scientists refer to as superorganisms), is equally beyond our grasp.

But the subtending relationship of the bacterial stratum to our own dominion goes beyond the challenges of negotiating coexistence in and through our mutually unfathomable natures. We have discovered enough about them to know that bacteria are the condition of our own possibility as multicellular beings: that they are at once our origin and our continuing vital support system. In an age of accelerating anthropogenic de-stratification, bacteria catch the fallout of our local and globalized transformations of earth systems, but we are the fallout of the dynamics of bacterial becomings (Hird 2009). We are the incidental inheritors of ancient bacterial symbioses and the recipients of the gifts of ceaseless microbial metabolism (ibid.). Whereas Latour (1988, 192) permits other actors to compose worlds of their own, the point about bacteria is their capacity to compose worlds for others – and by the same logic, their ability to withdraw or undermine the vital support that they provide for all other forms of life.

As such, the Anthropocene is a human signature, a superficial flourish, on what remains, indelibly, a bacterially orchestrated biosphere.

Even the most profound awareness of the intensified de-stratifications that are now surcharging the Earth's own de-stratifying tendencies should not detract from the profundity of our reliance on the world-making capacities of these other beings. In this way, even the notion of ecological interdependence may conceal as much as it discloses, for we are, above all, dependent. As is the case with other multicellular beings, the constitutive ecological transactions of human beings are profoundly and profusely asymmetrical (N. Clark 2011). Our existence is subtended and conditioned by bacterial life in ways that vastly outweigh the occasional localized dependence of bacteria on our offerings, an observation in keeping with Graham Harman's (2010a) insistence on the prevalence of asymmetrical relations in the known universe and on the corresponding rarity of truly symmetrical causalities (see chapter 8). By the same token, we ought to recall that Manuel De Landa's (1997) couching of a flat ontology, whereby all entities are equally agentic, some decades ago was entirely consistent with his exploration of a deeply stratified materiality. He posited a hierarchical structuring of the conditions of material existence that in no way implied a hierarchical valorization of all that exists.

Ontologically speaking, then, my point in sifting through the pits of accumulating human waste is less to highlight some grand anthropic rupture with the integrity of earth processes and more to prompt some sense of our inescapable, non-reckonable, and irrecompensable debt to other entities (Clark 2010). We may well spread our shit around to signal our possession of the spaces in which we dwell, as Michel Serres (2011) suggests. We may even inject our excrement deep in the earth to extend this stain into the layerings of geological time. Either way, what finally becomes of our defecations is up to the swarms of miniscule beings that ultimately engendered our existence.

CHAPTER EIGHT

# Wasting (in) the Anthropocene

## INTRODUCTION

The Anthropocene is a term that describes the impact that humans have had on the environment at a planetary scale. While the concept is already in popular circulation, its official acceptance as a geologic time or place is being considered by the Subcommission on Quaternary Stratigraphy's Anthropocene Working Group, which will make its recommendation to the International Commission on Stratigraphy (ICS) (Zalasiewicz et al. 2017). If accepted by the ICS, the Anthropocene will then be designated, physically, with a Global boundary Stratotype Section and Point (GSSP) – a literal golden spike to be placed somewhere on the planet; or symbolically as a Global Standard Stratigraphic Age (GSSA), which is a designated time boundary.

If stratigraphers decide that the Anthropocene marks a time boundary (after the Holocene), when would this time boundary begin? Some argue that it should be the late seventeenth century's Industrial Revolution, when the accelerated extraction and burning of fossil fuels began. Others place it some eight thousand years earlier in the Neolithic era, with the clearing of forests for agriculture. The even longer, deeper Anthropocene stakes a claim for the Promethean moment when humans learned how to harness fire, and put this knowledge to work through the widespread use of landscape burning. Until recently, Paul Crutzen (the scientist who revitalized the term Anthropocene) believed that the Anthropocene began with the large-scale extraction of fossil fuels, but he recently changed his mind. He now places the start of the Anthropocene on 16 July 1945

– the Trinity detonation, the first test of a nuclear device, and its radioactive waste fallout (Zalasiewicz and Williams 2015).

Or perhaps stratigraphers will opt to place a golden spike in a specific location. The Pleistocene-Holocene boundary – you can find it online – is in Greenland at 75.1000°N, 42.3200°W (Walker et al. 2009). In this case, there are any number of possible candidates: there is the Clarke Belt, approximately 35,786 kilometres above sea level, where orbiting satellite debris will outlast most, if not all, life on Earth; or anywhere within the 124,000 square kilometre range over which the Soviet satellite Cosmos 954 exploded in Canada's Northwest Territories in 1978, spreading some sixty-five kilograms of fissionable uranium-235.

Thus, whether it is the large-scale industrial use of fossil fuels (a form of necro-waste formed from the mainly anaerobic decomposition of buried dead organisms [Olson 2015]); the Trinity detonation that deposited radioactive waste into the stratosphere, or any of the subsequent nuclear accidents that added further radioactive debris into this stratospheric soup; or the ubiquitous dumps and landfills that proliferate the globe, archiving a timeline of extraction, consumption, and disposal – I make the case for *waste* as the signature of the Anthropocene. The question then becomes: How has waste inaugurated an epoch, one defined by the point at which human activity has intersected, in its significance and magnitude, with planetary geophysical forces? And, furthermore, what does this epochal wasting mean for accounts within Anthropocenic thought of a human claim of agency over geologic forces? What are the implications of defining humans' sum-total geological impact on the planet through our waste?

Geologists study the Earth as a vast "strata machine" whereby the Earth, "phoenix-like, has kept renewing itself" through a relentless process of vertical and horizontal sediment flows that move material – over billions of years – up through the biosphere and down through the lithosphere (Zalasiewicz 2008). And, as the previous chapter detailed, for the past four billion years or so – almost the entire history of the Earth – bacterial metabolic processes working in tandem with geologic forces have indelibly influenced (and often enacted) this de-stratification and re-stratification. The Earth, as Tyler Volk (2004) puts it, is "one big waste world" fuelling the world's organic and inorganic metabolism. Humans breathe tree and bacteria waste products (i.e., oxygen). We feed off the mineralogical by-products of the folding, eruptions, and implosions of the Earth's crust. The Earth's soil and water consist in large degree of the built-up waste products of living and nonliving matter (the mass of once-living organisms that have returned to base matter is estimated at somewhere between one thousand and ten thousand times the mass of the Earth itself (M. Davis 1996, 73). As such, waste is literally world-making, connecting geosphere, biosphere, and stratosphere (beyond the various gases and solid particles expelled as waste from organic and inorganic matter, myriad space junk orbits the earth). Human waste contributes to, and is ineradicably caught up in, the flux of earthly waste re-stratifications.

To wit, Canada's Giant Mine, located in the Northwest Territories, produced some 237,000 tons of arsenic trioxide dust – a highly toxic form of arsenic – as a by-product of gold mining.[1] The plan, so far, is to re-stratify the waste in the mine itself, using the frozen-block method to freeze this highly toxic waste in perpetuity for a future generation of engineers, scientists, politicians, and publics to resolve (see chapter 1). The plan for the Giant Mine, then, is effectively a permutation of our current dependence on landfills in which the intention is that these unfathomable tons of waste be kept in place: not leaking, not exploding, not combusting, not flowing beyond the boundaries we have imposed. We might also consider the Waste Isolation Pilot Plant about forty-two kilometres from Carlsbad, New Mexico, which is storing transuranic nuclear waste for at least the next million years. This site is already cut through with stratifications: the nuclear waste from hospitals, and plutonium from some 30,000 US nuclear weapons is stored in the hole created from the extraction of salt that was produced during the flooding and evaporation of the Permian Sea some 250 million years ago. This should lead us to consider the functions of a mundane dump or landfill – civilization's preferred method of disgorging its past.

We are refilling the Earth with materials we have already extracted, often in combinations that are highly toxic to humans, other animals, and the environment. In this way, the environmental crisis of the Anthropocene is a succinct summation of the empirically observable crises of waste: the wastes of modernity; of fossil fuels; of toxic and porous contaminated materials; along with pollution's contamination of bodies of water, the air, skin, shells, land, plants, and microbes. In understanding the Anthropocene as primarily an epoch of waste practices, attention must shift to the shadow economy (Dauvergne 2010), which haunts relational approaches to materiality: the exuberant, unforeseen, and often indeterminate consequences of material relations. This shadow economy, like waste, subtends relational approaches to materiality and has a determining and differentiating force in constituting the very possibilities of those relations. As the previous chapter detailed, as we manage our waste (i.e., as we sort, classify, and landfill our waste), microbes may well be generating new strata that in turn may well have consequences for humans and other living organisms.

As chapter 6 detailed, this bacterial waste-work is not without consequence. For reasons I explore later, bacterial mutation can be seen as a form of adaptation to the environment that at the same time generates new forms of life. These new forms of life may be generative of further new forms of life (and so on, into the future), which undercuts any human definitions of what "counts" as waste. Further, bacteria reveal abilities to create new life forms that challenge understandings of the relationality between humans and everything else. While focus remains on human waste as a stratification of the Anthropocene (retaining the anthropocentric focus of human/environment relations), bacteria are re-stratifying this waste stream with largely unknown results. As bacteria digest waste, they do not remain unchanged by this process of indigestion. Thus, empirical questions about microbial abilities to quickly reproduce, mutate, and potentially create new life forms raise ontological questions about what counts as waste matter, and how we should begin to apprehend the capacities of such matter to change, adapt, and diversify in evolutionary and materially inventive ways. This chapter argues that waste is transformative, not only of material conditions but also of the ontologies of matter and their theorization. To investigate this claim, let's look at the often obscured and indeterminate materialism that subtends biological and geological relations in the Anthropocene, to ask whether relationality is sufficient enough to encompass these speculative material futures and whether speculative thought is sufficiently engaged with matter's own speculative capacities.

## THE PROCLIVITIES OF WASTE

Far from the debris that lies forgotten in what we hope are contained and immobilized global dumps and landfills, waste flows.[2] That is, waste does not stay stratified within the social designations accorded it. Waste, as chapter 1 detailed, is a compendium of already extracted, assembled, reassembled, and transformed materials that, when dumped in a hole in the ground or modern landfill, is taken up – engaged – with leachate. So, all of those diapers, food scraps, metals, holiday wrapping paper, Styrofoam packages, pieces of wood, liquids, refrigerators, pets (as well as their shit and litter), batteries, chairs, pizza boxes, fabrics, and so on are re-stratified and compacted with other, rather less expected materials such as products of common industrial processes, like coal fly ash (of which over 50 per cent ends up landfilled; see Chertow 2009), plastics (more than 308 million tons of plastics are consumed worldwide each year, most of which still end up landfilled; see PlasticsEurope n.d.), and food waste, with its some 14,000 different kinds of additives and contaminants (over 97 per cent of which is landfilled in the United States; see Levis et al. 2010). To give one example, when food is recalled from Canadian supermarket shelves, it often ends up in landfills. XL Foods, Canada's largest food processor, processes over 40 per cent of the country's cattle and accounts for 30 per cent of the beef on store shelves. In the fall of 2012, approximately 5.5 million kilograms of beef presumed to be contaminated with *E. coli* was recalled, equivalent to 12,000 cattle. Of that, 500,000 kilograms were landfilled. When XL Foods wanted to reopen their plant to resume production, they were required to do a pilot test to ensure their corrective measures after the recall were effective. This test required the slaughter of 5,000 cattle, the carcasses of which were also landfilled after being tested for contamination, regardless of whether they had themselves been contaminated (Lougheed, Hird, and Rowe 2016; Lougheed 2017).

Landfills also mix hazardous and nonhazardous waste, including over seven million known chemicals, 80,000 of which are in commercial circulation (and with a further 1,000 new chemicals entering into commercial use each year; see Wynne 1987). This is not a stable heterogeneity over time; it is, in fact, dynamically reconstituting the terms of that heterogeneity, or as scientists put it, "variations [in leachate] may be cyclical, directional, stochastic, or chaotic" (Collins, Micheli, and Hartt 2000). Aerobic bacteria metabolize during the early life of a landfill, which produces material that is highly acidic and toxic to surface water. Anaerobic bacteria do the bulk of the metabolizing work deeper in the landfill's strata, producing leachate. Leachate is a heterogeneous amalgamation of heavy metals, endocrine-disrupting chemicals, phthalates, herbicides, pesticides, and various gases, including methane, carbon dioxide, carbon monoxide, hydrogen, oxygen, nitrogen, and hydrogen sulphide. Factors affecting leachate production rate and composition include the:

> characteristics of the waste (initial composition, particle size, density and so on), the interaction between the percolating landfill moisture and the waste, the hydrology and climate of the site, the landfill design and the operational variables, microbial processes taking place during the stabilization of the waste, and the stage of the landfill stabilization. Most of these factors change during the operational period of the landfill as the landfill is developed causing significant changes in leachate quality and quantity. (Yildiz and Rowe 2004, 78)

Leachate may travel vertically and horizontally within landfills and may continue to travel when it leaks beyond landfill cells, sometimes through geological strata. That is, leachate may percolate into soil and groundwater, where it moves into and through plants, trees, animals, fungi, insects, and the atmosphere. Via leachate, bacteria create well known, little known, and new biological forms. Greg Kennedy notes, "Trash may dissemble the truth of its being by presenting itself as [an] immaterial, innocuous substance divorced from the relations to physicality," but in actuality, biogeological processes are always – and already – actively involved (2007, 162). The millions of people who live in and on dumpsites, and who survive by directly handling waste, or who live downstream from waste's fallout, are acutely aware of waste's proclivity to flow (McGovern 1995; Lepawsky and Mather 2011).

## THE RELATIONALITY OF WASTE

Karen Barad's theory of agential realism offers a rich articulation of relationality, and is inspiring new ways of engaging in philosophical discussions about longstanding concerns with ontology and epistemology, naturalism, realism, and constructivism, and with more recent calls for interdisciplinary modes of theoretical and empirical engagement. Deploying Barad's theory within waste studies, we may recognize that waste is less a "thing" or even a distributed set of things. On the contrary, waste requires material-discursive constructions that create, bring to the fore, and sustain particular relations and that deaden, obfuscate, and otherwise limit other relations. As such, waste has a genealogical trajectory that subtends and exceeds objectification. From an agential realist perspective, the world is not composed of entities that we may or may not come to know; it is composed of "phenomena" produced through measuring. Agential realism's focus on phenomena rather than things requires careful attention to the myriad "material-discursive practices" through which these phenomena are (re)constituted, (re)arranged, and abandoned (Barad 2003, 818).

Agential realism provides a theoretical framework within which to analyze the various material-discursive "apparatuses" that "(re)configure" and literally make and remake the world's waste (Barad 2003, 816). Within this theoretical formulation, then, waste is not a given thing that awaits analysis and critique (as well as sorting, reclaiming, and disposing). Waste is a phenomenon through various agential "cuts," or what Barad calls "cutting together-apart" (2012a, 14). These cuts require various apparatuses and practices such as mass-production and consumption, anaerobic digestion, global transportation and communication, statistical modelling, organismal differentiation, cheap mechanized labour, heavy water, cultural analysis, non-renewable fossil fuels, regulations, and so on. For example, the water safety standards through which our government reassures us that the water (whether tap, lake, stream, pond, or bottled) is safe to drink involve a complex "cutting-together-apart" of scientific studies (themselves involving many cuts – see Hird 2012), committee deliberation, industry lobbying, bacterial reproduction, contaminant movement through human bodies, and so on. Measuring drinking water safety, or knowing whether drinking water is safe, is "world-making." Measuring "entangles"; or as Barad puts it, measuring "cuts together-apart" (Barad 2012a, 14). From this onto-epistemological perspective, matter and meaning are measuring "effects" (ibid., 14). They do not pre-exist as individual entities, nor are they inherently static in time or space (to make them so is itself to exact an agential determination, or "cut," as Barad puts it (ibid.).

When we focus on things, we obscure the primacy of phenomena as "dynamic topological reconfigurings/entanglements/relationalities/(re)articulations" (Barad 2003, 818). So, when we analyze waste as a thing, it risks appearing as rather static, like the garbage dumped in landfills or the plutonium buried in nuclear repositories, as though materiality is just waiting there, passively, not doing much. But waste is anything but static and submissive: waste flows and mobilizes relations (Gille 2010); waste enacts, as chapter 7 argued, new material phenomena; and waste forces new social and geological formations into being. The unfathomably diverse and multitudinous bacteria that metabolize landfills, and the half-lives of iodine, strontium-90, plutonium and other nuclear materials, suggest just the kind of (re)configurings, entanglements, and relationalities that agential realism forefronts.

This is not to say that agential realism denies the existence of objects outside of or prior to relations. As Barad writes:

> properties that we measure are not attributable to independent objects. Independent objects are abstract notions. This is the wrong objective referent. The actual objective referent is the phenomenon – the intra-action of what we call the electron and the apparatus. (2007, 61)

Although we do not have access to objects in themselves, we may gain some insight into what agential realism refers to as phenomena or what speculative realism details as the types of relations objects may have (containment, contiguity, sincerity, connection, or none; see Harman 2011b), and how these relations affect objects.

What makes agential realism particularly relevant to waste studies is that this theory forefronts the inhuman in measuring/knowing. Barad's work reminds us that world-making is not affected through human measuring and knowing alone. "The world," Barad reminds us, "theorizes as well as experiments with itself" (2012a, 2). Barad, turning human exceptionalism on its head, argues, "If we thought the serious challenge, the really hard work, was taking account of constitutive exclusions, perhaps this awakening to the infinity of constitutive inclusions, the in/determinacy that manifests as virtuality calls us to a new sensibility" (ibid., 13).

## A PLATE TECTONICS OF ONTOLOGY

This attention to the inhuman is taken up within the expanding swathe of theories called speculative realism, object-oriented philosophy, object-oriented ontology, and their variations. A central concern uniting these theories on metaphysics – whether Ray Brassier's eliminative nihilism, Iain Hamilton Grant's cyber-vitalism, Quentin Meillassoux's speculative materialism, or Graham Harman's speculative realism – is an objection to Immanuel Kant's settlement, or what Meillassoux defines as "correlationism": "the idea according to which we only ever have access to the correlation between thinking and being, and never to either term considered apart from the other" (2008, 5). "Every philosophy," writes Meillassoux, "which disavows naïve realism has become a variant of correlationism" (ibid.). This includes, then, all variants of phenomenology, structuralism, and poststructuralism, and certainly postmodernism. Speculative realism takes issue with phenomenology's insistence that we cannot know reality itself, but only objects as they appear to us (through experience, scientific experimentation, modelling, traditional ecological knowledge, and so on). It also takes issue with those formulations of science studies such as Bruno Latour's actor–network theory for their tendency to reiterate the salience of human consciousness within analyses whose ostensible focus is objects themselves. As Harman observes:

> [Latour's] examples are drawn from the human realm, not from general cosmology. And in this way, the more difficult cases are left in shadow. With a bit of work, it is not difficult to see why all objects that enter human awareness must be hybrids, why the ozone hole or dolphins or rivers cannot be viewed as pure pieces of nature aloof from any hybridizing networks. The harder cases involve those distant objects in which human awareness is currently not a factor at all. Where are the hybrids in distant galaxies? If they are not present, then the purifying discourse of nature wins the war, and the rule of hybrids can be viewed to some extent as a local effect of human perception. (2008, 8)

Speculative realism, for Harman, attempts to prioritize objects themselves, rather than their relations with humans, or, indeed, other objects. Objects, within Harman's formulation, are somewhere between the tendencies to either "undermine" or "overmine" within philosophy. When objects are undermined, a form of recursive monism occurs, as what appear to be autonomous objects turn out to be aggregates of smaller objects. If these objects are not distinct from one another, then we have monism; if they are distinct, then their relation to each other is merely an assertion that objects are "both connected and unconnected at the same time" (Harman 2011b, 9). This assertion may come in the form of popular tropes of "becoming" or "difference" but leaves unaddressed the issue of the object itself – if each becoming or difference is merely an instance of one underlying becoming or difference, then we are back to monism, and if each is distinct, then we have objects whose becoming or difference from other objects remains to be explained. When objects are "overmined," argues Harman, their only existence consists of their relation with other things. Here the object itself recedes as relationality takes precedence. To be sure, this may not need to be relations between humans and objects, but we are left with the same issue we have with undermining: relationality is abstracted such that objects only count as such when they affect other objects (objects are entirely exhausted by this relationality – nothing is kept in reserve), which forecloses the possibility of change.

Working from the rich philosophical lineage of Heidegger, Whitehead, Lingis, Husserl, and others, Harman's object-oriented philosophy is at pains to separate object from subject. Indeed, Harman defines an object as "anything that has a unified reality that is autonomous from its wider context and also from its own pieces" (2011b, 116). Or as geographer Ian Shaw puts it, "objects reduce each other to caricatures – they literally 'objectify' each other" (2012, 620). Instead, relations "do not exhaust" the things that relate; objects are more than the sum of the relations and networks within which they (at times) participate (Harman 2011b). Thus, whereas in some relational formulations such as Latour's actor network theory, black-boxing is something for social scientists to work on, for Harman it is something objects themselves do.[3]

Objects, within Harman's metaphysics, do not have to be physical things; they may equally be deities, dreams, faeries, or fantasies. Thus, all objects have hidden qualities, and may also be dormant for short or eternally long periods of time. In an oft-cited illustration, Harman writes:

> When fire burns cotton, it makes contact only with the flammability of this material. Presumably fire does not interact at all with the cotton's odour or colour, which is relevant only to creatures equipped with sense ... The being of cotton withdraws from the flames, even if it is consumed and destroyed. (2011b, 44)

Unable to fully touch or otherwise experience other objects, objects affect each other through what Harman terms "vicarious causation" (2007). Two or more objects can only influence each other by forming, as it were, a third object, which becomes "the molten inner core of objects – a sort of plate tectonics of ontology" (ibid., 174). Given that neither of these objects needs to be human (or indeed within the scale of the universe are likely to be human), vicarious causation holds that relationality itself (and not human consciousness) necessarily and always distorts the reality of objects. Being in relation, within Harman's speculative realism, does not refer to the essence of any object in itself, but only to a part of that object (which Harman refers to as its "sensual" element). This determination of relationality as distortion fits well within agential realism's proposal that phenomena are produced through intra-active cutting-together-apart.

## THE SPECULATIVE POSSIBILITIES OF WASTE

Science and technology studies bring into sharp relief the processes through which things become objects, objects become black-boxed, and humans enact systems of classification and other practices such that objects appear to be stable or unstable, fixed or transitory, hidden or exposed. What brings agential and speculative realisms together is, as I have outlined, a concern with the inhuman, and more specifically with figuring out ways to explore a universe that does not require human consciousness to either exist or be understood. Both agential and speculative forms of realism ask us to contemplate a world of objects unmediated by human consciousness, one emphasizing the "cutting-together-apart" that all objects are (human or otherwise), and the other emphasizing the withdrawn, unavailable aspects that maintain object autonomy. Within both approaches, we get a keen appreciation of the unending dynamism of objects and their relations: that there is no relation or object that is so durable that it cannot be worn out, overthrown, bribed, caressed, or otherwise moved by its own will towards something else. At the same time, an object is not simply the sum of its relations, and can withhold, suspend, or reserve forms of relation. For Shaw, a "geo-event" occurs when these transcendental objects – objects that affect all of the objects in a world – "are overthrown by inexistent objects," thereby bringing forth a new constellation of objects, a new world – worlds brought forth by bacterial-oriented objectification in the Anthropocene, for example. This is made possible because no object is exhausted by its relations with other objects. There is always a surplus. As Shaw explains, "object by object, the geo-event de-anchors the integrity of the world" (2012, 622). If, indeed, the Anthropocene is a geo-event of epochal magnitude, there is much need to address not just this exhaustion and survival of objects (species, biodiversity, minerals, fuels, humanity) but of the inexistent objects that both form the shadow economy of these social relations and exceed them.

What also unites these metaphysics is a strong recognition of indeterminacy. Within speculative realism, vicarious causation maintains that all relations must, in the relating objects' refusal to reveal all of their qualities, be indeterminate. When objects form relations with other objects, the withdrawn aspect(s) of each object means that each relation can only be partial and never entirely known or certain. For agential realism, indeterminacy is the condition of measuring: "how strange," writes Barad, "that indeterminacy, in its infinite openness, is the condition for the possibility of all structures in their dynamically reconfiguring in/stabilities" (2012b, 16). Astrid Schrader describes the articulation of the trace, when indeterminacies are rendered determinate phenomena through measuring:

> There is no measurement without memory, no intra-action that wouldn't leave a trace. But the trace by itself is not. Memories have to be read in order to "be." That is, they require work that involves material determinations. (2012, 43)

The work Schrader refers to is determining phenomena: the work that comprises knowing. The trace, then, is a provocation, and knowledge is a form of cutting. Moreover, we cannot, writes Schrader, both be part of nature and examine nature from an external position. It is not that we are part of a system of which particular parts are currently inaccessible though accessible in some future (Heisenberg's uncertainty), but rather "because the scientist's work helps to enact the system boundaries she will have become part of" (Schrader 2012, 44). Knowing, then, is about determining the indeterminate. It is about setting limits; cutting together-apart (think of drinking water safety standards, air quality standards, and so on). This is a limitless process; so long as there are phenomena, there is indeterminacy rendered determinate:

Matter is never a settled matter. It is always already radically open. Closure can't be secured when the conditions of im/possibilities and lived indeterminacies are integral, not supplementary, to what matter is. (Barad 2012b, 16)

Thinking with indeterminacy as an integral condition of matter provides a much more expansive mode of recognition of the speculative possibilities of objects than is admitted within an anthropocentric view of the Anthropocene and its waste-worlding. Recalling Barad's provocation that the world experiments with itself, we have little idea what the multitudes of bacteria that metabolize landfills will make of its ingredients, of themselves as they proliferate and differentiate into new forms, or of the geosphere and biosphere, given enough time (see chapter 7).[4] The issue is that the radical openness of matter does not cease being open to various forms of experimentation when it is designated as waste or put underground. Bacteria continue to metabolize the various assortments of left-over materials that are aggregated together in a landfill in ways that defy their given stratification as waste, enacting new flows that exceed this physical containment. More than the Anthropocene's discontents, waste sites cannot simply be understood as a collective unconscious of industrialization that will return, like the repressed, to offer a critique of the psychic life of wasting.

Civil engineers and scientists are tasked with technically managing waste within the context of indeterminacy. For civil engineers, WM is all about the task of determination and containment. Determining waste is a highly complex process, and the constitutive inclusions are extensive. Landfills, nuclear repositories, incinerators, and other management techniques have become more technically sophisticated as engineers and scientists develop better liners, gradient specifications, barriers, and so on (though it does not mean these techniques are always adopted). These management techniques are developed through conceptual and statistical models (of, for instance, the movement of moisture, bacterial metabolism, and soil integrity) and then through testing these models empirically in controlled experiments. For example, the assumption in conceptual models constructed to test landfill quality is that each layer is a completely mixed reactor with uniformly distributed waste, moisture, gases, and bacteria – but this is not actually the case, with cells containing decomposed material of varying degrees (Yildiz, Ünlü, and Rowe 2004). In other words, engineers and scientists are knowingly creating and operating within conceptual models (rather than reality) in order to derive determinate matters of fact. Describing the principle of "renormalization" in physics, Barad writes:

> If it turns out to be possible to get finite results by subtracting infinities via a process that cuts out the domain of unknown physics, then the theory is said to be re-normalisable. The cut-off method of renormalisation is a mathematical way of bracketing out what you don't know. (2012a, 11)

Engineering and science derive certain facts about landfills, bioreactors, and the like that are known through this necessary process of bracketing out, or subtracting, indeterminacy – not just what is as yet unknown, but what is unknowable. Engineers and scientists are well aware of indeterminacy in, for example, their attention to issues such as contaminants of emerging concern, which include chemicals such as bisphenol A (BPA) that have been used in many plastic products and are believed to mimic human estrogen at low concentrations (LaPense et al. 2009; Takai et al. 2000), and polybrominated diphenyl ether (BPDE), which is an additive flame retardant in plastics, foams, and fabrics that may cause liver, thyroid, and neurodevelopmental toxicity – as well as new materials such as nanoparticles, which were not part of the waste stream at the time many landfill regulations were developed (such as US Subtitle D) (Islam and Rowe 2009; Rowe 2012; LaPense et al. 2009; Takai et al. 2000). Issues such as contaminants of emerging concern are not just "known knowns" (the things we know we know), "known unknowns" (the things we know we don't know), or "unknown unknowns" (the things we don't know we don't know) (Rumsfeld 2002): the words "contaminants," "emerging," and "concern" are themselves agential cuts. Suspending the connotative powers of language in bracketing off indeterminacy is a first step towards the kind of disruption that might allow another economy of relations to be disclosed – one that keeps waste's own indeterminations as a concurrent concern. The idea is that for every cut of knowledge there is a corresponding memory of what is withdrawn in both the process of determination and that account of matter. In a somewhat strange philosophical coupling, Barad's theorization of indeterminacy together with Harman's understanding of the object's withdrawal provides a way to articulate a form of material speculation – which may or may not be called the Anthropocene – that is neither wholly determined by human relations, nor immune to the political accounts of matter that seek to determine it.

## UNDERMINING WASTE

If one of the concerns of the Anthropocene is its articulation of humans' de-stratification of billions-year-old fossil fuel and other material layerings from the earth's strata, it also articulates an opposite concern, that of landfills and nuclear waste repositories as a kind of earthly re-stratification or re-layering (see chapter 7 and the epilogue). A landfill's contaminating lifespan is estimated at hundreds to thousands of years, and nuclear contamination endures for upwards of 100,000 years, or 3,000 generations, making the consequences of this re-stratification indeterminable. Such de-stratifications and re-stratifications of the Earth's geologic layers have consequences, as new formations of matter are brought together with little or no knowledge of how these cuts continue to germinate (or evolve) into new historic strata. While waste may be often categorized as the end life of an object and its material utility, it is the starting point of a new form of material ontological speculation.

Not only are these wastes materially indeterminable, but their duration promises for themselves much more of a future than the one that humanity may wish to lay claim to. The geologic-like cut in time that occurs through waste practices establishes something akin to what Meillassoux has called "ancestral statements" (albeit from the future anterior rather than the past) (2008, 10). Meillassoux makes the distinction between primary and secondary qualities in objects, primary qualities being non-relational and secondary properties being relationality enacted. Primary qualities are not available to enter into relations (because they either precede or exceed human consciousness) and can only be inferred through the fossils that exist in the present. He defines ancestral statements as events that are anterior to the emergence of all conscious apprehension and therefore have no body of affects or perceptions to register their event. This matter is not a fossil in the traditional sense of being a trace of life, but rather an *arche*-fossil that indicates the existence of realities anterior to all life, such as a radioactive isotope. Meillassoux's point is that we can have knowledge of non-relational properties in so much as they pre-exist us, or in the case of waste, will certainly outlive us. If we view waste as a ruin in reverse, producing its own futurity or deep time, independent of but not unconnected to human futures, this makes room for thinking about the ontological junctures of determining indetermination that lie beyond what determination means for us.

From a waste studies perspective, an important contribution to contemporary studies of the Anthropocene involves attending to the ways in which humans create waste from objects, then attempt to divest themselves of these objects through various forms of disposal (whether in underground nuclear lairs, landfills, dumps, energy-from-waste [EfW] facilities, and so on), and then leave engineers and scientists with the task of dealing with this dispossession's remainder – fly and bottom ash, leachate, contamination, and so on – especially when this divestiture is considered within the context of the already dispossessed, human or otherwise. Waste in this sense is the cutting-together-apart of a complex array of objects, measuring practices, and humans.

At the same time, we might also consider that real objects are withdrawn from our apprehension. Bacteria, for instance, in their awkward and stubborn refusal to be easily characterized as a species, and in their sheer numbers and diversity, certainly defy – and indeed are largely indifferent to – the claiming and naming practices of humans (Hird 2009). Yet, it is the very relations that bacteria form with each other, and with moisture, soil, sunlight, various chemicals, and so on within landfills, that makes the phenomenon of landfill waste what it is. Similarly, what makes radioactive uranium is its relation with time and organic bodies. In other words, leachate, as an object, is already a network. It is difficult, by definition, to separate it from the masses of objects, moisture, and teeming bacteria that make it up. We are left with, as Mattias Kärrholm (2013) points out, a relational typology.

Moreover, one of the things that characterizes scientific and applied scientific approaches to forms of waste (radioactive material, leachate, fly ash, and so on) is precisely the attempt to disaggregate (to undermine, as Harman would have it) these objects into their constituent parts. It is difficult to examine any one constituent of leachate, let alone one constituent's synthesis with another or with several others. As waste becomes gathered up into the term "leachate," a further mode of differentiation is made between matter that stays stratified and matter that flows. We may see waste as this double articulation (in Deleuze's terms) of the unstable forms between strata – waste being an inter-stratum between social and geophysical processes that regulates relations between strata. Gilles Deleuze and Félix Guattari refer to a "double articulation" that is meant to take account of this constitution of entities (1980, 39). This is not done, as Aristotle described it, through determining essence, but rather, first, through the selection or sedimentation of materials that will make up the object, and, second, through the further sedimentation of these materials into an entity with its own properties. For leachate to be such an object, there must be an ongoing process of extraction, sedimentation, re-sedimentation, and so on. What marks these stratifications of waste is the mixing of wildly different strata with no single plane of consistency, such that it is not enough to merely confirm the exclusion of the indeterminate without also acknowledging that the indeterminate determines future possibilities, even as these remain anterior to human life. Landfills and other forms of waste may have the appearance of sedimentation, but this is in human time, and sometimes not even then. Such mixing of strata is a new form of engaging in geology with the planet, in so much as it risks on the one hand a wild and furious de-stratification of the earth (of the carbon, nitrogen, and atmospheric and ocean cycles) and on the other introducing unknown stratifications of indeterminate parts that materially refuse to stay stratified.

If waste phenomena (that of humans, strontium-91, landfills, recycling, non-renewable fossil fuels, anaerobic bacteria, and so on, as Karen Barad's theory would formulate) are always already relational and marked as distinct through ongoing processes of cutting-together-apart, what might this add to, challenge, or otherwise provoke in discussions about the politics and ethics of waste? And how might this working through of the cuts of relationality prompt us to think about how matter exists independently of humans but also exists together with our own waste-making practices? Examining waste as a speculative material practice involved in processes of stratification requires an understanding of the interpolation of epistemological and ontological terms in acts of cutting by both humans and inhuman organisms. While managing waste lends a spatial and temporal specificity to these practices in the short term, waste's indeterminate futurity suggests that waste practices need to include a consideration and understanding of waste beyond its relationship with humankind. If, as Foucault reflected, "knowledge is not made for understanding; it is made for cutting" (Foucault 1984a, 88), then an acknowledgment of the speculative dimensions of the significance of waste beyond human determinations creates different sets of ontological relations to waste that are non-relational, in the sense of being anterior to the present and independent of human attempts to designate waste as *waste*. Through the transformative capacities of waste, then, we are able to comprehend an inaccessible deep time in which the human subject is necessarily discarded as an object of primary concern. Thinking about the speculative possibilities of bacterial waste-life inadvertently lays open political and material possibilities in processes of (in)determination that might transform what these processes become.

Ultimately, Barad's agential realism is a theory about ethics: it seeks to address the difficult ground of social obligation through (re)articulations of phenomena, which, to use Isabelle Stengers's words "allow us to inherit our history otherwise" (in Bordeleau 2011, 17). It also requires, even demands (in a Derridean poststructuralist sense), that we pay attention to the indeterminate inheritances that compose and foreshadow our heavy responsibility for the future to come (Diprose 2006). While the concerns of these all too human ethical relations may seem at odds with some of the claims made in this chapter with regards to inhuman agency, it is precisely the mode in which matter is politicized through ontological inquiry that informs processes of (in)determination. Barad's contribution is to take these ethical concerns through a formal engagement with matter across the field of human and inhuman cuts, while retaining a hold on the agential specificity of those cuts. In dual concern for responsibility to both matter and its determinations, there is in this process a sympathetic taking up of the obligations to the future that motivated some poststructuralist thought, alongside an understanding of inhuman agency within, and on the outer reaches of, that thought.

The indeterminacy of waste draws our attention to the imprescriptibility of our ethical responsibility to future generations and environmental sustainability (Arendt 1958; Benford 1999; van Wyck 2005). In law, prescription infers a statute of limitations on identifying and assigning guilt: after a certain time, crimes must be forgiven. Imprescriptibility refracts a global futurity of indeterminate entanglements, of cuttings together-apart, and of collective (in)human vulnerability and responsibility:

ethics is not simply about responsible actions in relation to human experiences of the world; rather, it is a question of material entanglements and how each intra-action matters in the reconfiguring of these entanglements, that is, it is a matter of the ethical call that is embodied in the very worlding of the world. (Barad 2007, 160)

Waste, then, draws our attention to the complex de-stratifications and re-stratifications, the limitless potential intra-actions corralled into composing waste as human, inhuman, disposable, reusable, risky, determinate, containable, profitable, inert, anthropogenic, and ethical.

## WASTE AND THE ANTHROPOCENE

So where does the agential realism of waste lead us? Is viewing waste as a semi-autonomous object that generates its own futures, as Nils Johansson and Jonathan Metzger fear, to somehow treat it as pre-ordained or given (2014, 4)? If an acknowledgment of the futurity of waste, independent of human concerns, provides too much of an ethical alibi by disowning the inheritances that are clearly part of industrial relations, then there is at least a need to retain a dual aspect of objectification (as both inheritance and material speculation). Speculation then plays an important role by luring us into the existences of those dual aspects, not to resolve that indeterminacy, but to partake of its inaccessibility (Yusoff 2013, 211). Attention to that which is insensible or anterior to us is part of noticing how determination cuts out that which "troubles our sensibility/sense-of-ability as well as limiting our comprehension and compassion for that which is outside of our sphere of experience" (ibid.).

How matter is made to appear, and the ontological and epistemological technologies of recognition that provide a certainty to its appearance, are clearly part of the politics of waste, but it is but one part of the story. What becomes crucial in these ontological politics is how we both speculate about the agency of indeterminate matter and account for the speculative capacities of that matter to become otherwise, along radically different spatial and temporal axes. This is perhaps the lure of the conceptualization of the withdrawn in Harman's object-oriented philosophy. What objects appear to be – their aesthetic and sensible allure – is a matter of politics, not all of which is human (De Landa 1997; Hird 2009). Ideally, we want to bear in mind the constant worlding taking place that is insensible to us, rather than assume that all that matters is what affects us and provokes us to notice it. Furthermore, the topology of effects that waste might generate is simultaneously happening on a much different temporal and different spatial configuration than is often given through the determining cuts of policy legislation and waste-practice management.

Landfills and other underground waste burial sites are particularly illuminating examples of other-worlding. The dark biosphere refers to the ecosystem at least one metre below the continental surface or seafloor. Up to 95 per cent of bacteria and archaea live in this deep subsurface (Edwards, Becker, and Colwell 2012) and exist independent of light as a source of energy. These lithotrophic microbes rely on various gases (such as carbon dioxide and hydrogen) for energy to do their metabolic work. That we know so little about this dark biosphere, literally beneath our feet, may well be one of the biggest blind spots in Anthropocenic practices. Grappling with the indeterminacies of this dark substratum requires searching out modes of recognition that do not depend upon our usual human sensorium, both ontologically and through our various epistemic cutting tools. If "existence is coexistence, as it were, through mutual abandonment and exposure rather than enclosure" (Yusoff 2013, 220), then the containment of waste as a category of disappearance in Anthropocenic thought needs to be thrown off. As Kevin Hetherington argues, "disposal is about placing absences and this has consequences for how we think about 'social relations'" (2004, 159). On one level, the formulation of waste *as waste* obscures the very fact that it is the disappearance of these objects from view that renders the ontology of these objects indeterminate (as well as rendering the work of determination doubly problematic, because it is proceeded by human indeterminations and exceeded by bacterial determinations). Both actions (human and inhuman) collude to generate indeterminate objects. A politics of recognition, then, is not the point: a careful reading of the speculative possibilities of what waste might become through its own inventive capacities, with and without a human subject, quite possibly is.

CHAPTER NINE

# The Indeterminate Material Politics of Waste

## INTRODUCTION

As the previous two chapters detail, the concept of an Anthropocene captures an emerging recognition and interest in, simultaneously, the operation of human-created infrastructures and global politico-economic practices characteristic of industrial capitalism, and geological and biological processes stretching back through deep time. Whether in the form of mining, nuclear, industrial, hazardous, sewage, or municipal, and whether it is dumped, landfilled, incinerated, or buried deep underground, waste constitutes what will likely be the most abundant and enduring trace of the human species for epochs to come. While stratigraphers debate the appropriate geographic coordinates for the next GSSP golden spike, or where to mark the temporal beginning of this geological era, the real provocation of the Anthropocene is not that we are leaving a message for some imagined future humanity to decipher, but that we are bequeathing a particular indeterminate futurity through a projected responsibility to our children and all those children to come, a theme I will return to in the epilogue.

This chapter synthesizes the ideas in the previous two chapters concerning waste and the Anthropocene, the exploration in section 2 of Canadian settler colonialism and waste, and the focus in section 1 on how Canadian waste is governed by industry and government, in order to reflect on the argument posed by Dipesh Chakrabarty (2013) that the settler colonial subject must be conceptualized within the context of the Anthropocene. The Anthropocene, Chakrabarty argues, re-characterizes the settler colonial survivor as both the colonized and – as a geological force – the re-Indigenized, insofar as humans may no longer think of themselves as separate from nature. That is, within a particular logic of Anthropocentic discourse, Chakrabarty argues, the colonized subject may well end up subsumed within a universalized *homo sapiens*; a species for whom, in functioning as a geophysical force, sovereignty is no longer possible. He explains:

> It has to be one of the profoundest ironies of our modern history that increasing use of such energy [fossil fuels] should have now transformed our collective image, in our own eyes, from that of an autonomous if not sovereign and purposeful agency – from the level of individuals to the level of groups – to that of a force, which is defined as "the sheer capacity to produce pull or push on an object by interacting on it merely as another object." In other words when we say we are acting like a force, we say we don't have any sovereignty. We are like another object. A geophysical force has no sense of purpose or sovereignty. (Ibid., 1)

Thus, on one hand, the Anthropocene, as a discourse, may well become a universal *de-*colonizing project insofar as it challenges humanity's separation from, and superiority over, nature. On the other hand, a humanity based on a universalized postcoloniality erases Indigenous history, as well as Indigenous ways of knowing and being, in favour of a universal human being with globalized technologies – geoengineering and big science. Chakrabarty's characterization of sovereignty is based on a knowable, stable, and predictable concept of geology – one that Indigenous scholars argue has never existed, and could never exist (Cameron, Mearns, and McGrath 2015; Qitsualik-Tinsey 2015). Inuit did not sustain an "anthropogenic" sovereignty prior to colonization, yet Inuit struggling for self-determination are required to adopt this framework in negotiations with Canada's federal government (Qitsualik-Tinsey 2015, 27; Price 2007). Techno-managerial approaches to the ubiquitous, toxic, and indestructible wastes of the Anthropocene hinge on a sovereign approach to human/nature relations. Recall the discussion in chapter 4 of the ideology that legitimated the colonization of Inuit, First Nations, and Métis peoples in Canada. This ideology was predicated on the ontological separation of (colonizing) humans from nature. Recall as well from section 2 that the waste in Canada's

Figure 9.1 | *Taima* (Stop). Julie Alivaktuk in front of the West 40 dump in Iqaluit, Nunavut.

Arctic (both its volume and toxicity) is a settler colonial dividend. US and Canadian militaries, as well as American and Canadian prospectors, industry, and settlers all introduced waste to the Arctic, and – for the most part – abandoned it there.

The dangerous irony of the Anthropocene, then, is less that the possibility of sovereignty has collapsed, and more that the various technologies that many hope will solve our global environmental problems are framed through an understanding of sovereignty that always separates waste from resource, dirt from cleanliness, and uncivilized from civilized – a configuration that, as the Anthropocene has already begun to prove, is doomed to failure. This past, and the future promise of waste – the environmental fallout of increased oil, gas, and mineral extraction; military installations; shipping; and the tourist trade to the health of humans and inhumans – projects a responsibility for present and future generations to resolve. The uneven distribution of both the causes of Anthropogenic change and the effects of this waste landscape – capitalism's implicit dividend – are differentially experienced by Inuit in the Arctic.

## DUMPCANO

On 20 May 2014, Iqaluit became the subject of controversy when its West 40 dump (see chapter 5) spontaneously caught fire for the fourth time in under a year. Within twenty-four hours, the fire had spread throughout the city's four-storey football field–sized open waste disposal site.[1] The Iqaluit Fire Department's initial attempts to extinguish the fire with water failed, and fire crews moderated their strategy to containing the fire by digging additional fire breaks, including a trench between the smouldering section of the dump and a smaller older pile, which constituted "the most hazardous part of the landfill," according to the fire chief (Varga 2014d, 1). This technique essentially involved submerging burning waste into a lined containment pool while simultaneously spraying the dump with sea water sourced from Frobisher Bay, and recycled leachate.

Government officials, including Environment Canada and Health Canada representatives, released a series of public health warnings directed at an alarming number of vulnerable residents: children, elderly people, people with respiratory issues, pregnant women, women who might become pregnant, and women who might not yet be aware that they were pregnant. Yet, these warnings were issued alongside official reassurances that contamination levels were not a cause for concern. Local residents largely rejected the government's messaging and created their own counter-messaging, which included various activist interventions, such as employers distributing gas masks to their employees, local school administrators cancelling children's classes, an activist group organizing an art installation about the dump fire, and a lively Twitter feed that included a photograph of a pregnant Inuk woman wearing a mask and with the word *Taima* (Inuktitut for "enough") written on her hand.

After more than three and a half months, the fire was finally extinguished. The Canadian federal government, claiming that the dump fire did not constitute an emergency, declined to provide the funds needed to either extinguish the fire or renovate the dump into a modern landfill resembling those found in southern Canadian cities. This abrogation of responsibility left it to the City of Iqaluit to absorb the financial burden of managing and extinguishing the fire through its city budget. Local residents – both Inuit and *Qallunaat*[2] – claimed that the dump's constant smouldering and sporadic combustion were not just a symptom of a technological or political failure to extinguish the fire and deal with the health effects of the dump smoke lingering over the city and the leachate running into the nearby estuary, but also a symptom of an ongoing settler colonial legacy that sustains a two-tier system favouring the health and welfare of southern Canadians. For their part, southern government officials and scientists focused on both the limitations of contaminant measuring and monitoring, and what they perceived as the public's deficiencies in understanding and acquiescing to official messaging that there were no significant threats to human health or the environment.

This chapter brings together all of the arguments developed so far in this book. I am interested in the ways in which the indeterminate materiality of the dump – the mixture of materials, atmosphere, smoke, bacteria, leachate, and so on – became entangled with concerned citizens, public service announcements, and municipal budgets, as well as discourses related to health and safety, sovereignty, fiscal and environmental responsibility, and, implicitly, ongoing settler colonialism and consumer responsibility. That is, the West 40 site became the controversial "dumpcano" when its materiality (or more specifically, its multispecies (micro)biology and geology) became entangled with a number of economic, cultural, political, and social conditions, enabling the indeterminacy of waste to come to the fore.

As I argued in section 1, waste in southern parts of Canada is largely configured as a techno-scientific issue in which waste is removed from curbsides and mainly deposited in landfills that are out of sight and out of mind. People are largely classified as consumers, and as consumers, are held responsible for generating waste, post-consumption. The management of waste is typically so routinized in southern regions of Canada that – aside from when there are changes or proposed changes to the management of waste – it does not garner or sustain the public's attention as an issue, and thus it does not become political. Even in Canada's Arctic communities, where waste is left on the land close to communities at open dumpsites, or dumped in the sea (a chronic problem with cruise ships that peruse Arctic waters), the lack of funding, infrastructure, and qualified personnel to construct, maintain, and provide aftercare for modern landfills, incinerators, or EfW facilities, coupled with other extraordinarily urgent economic, infrastructure, environmental, and social issues, means that waste in the Arctic – while certainly not out of sight, nor out of mind – is forced to take its place amongst myriad other issues. The West 40 dump fire that caught the attention of southern Canadians through various mainstream media (CBC News and the Canadian Press, for instance) brought the mundanity of WM into the spotlight and transformed waste from an object into an issue (see chapter 2). For local residents of Iqaluit, and for people living in other Arctic communities, dump fires have been an ongoing issue since settler colonization, and dump fires in Iqaluit and other communities such as Rankin Inlet are regular features in the *Nunatsiaq News*.

Drawing upon archival sources and primary interview data that my graduate student Alex Zahara (2015) gathered in Iqaluit (I was able to complete some interviews as well), this chapter will examine how officials used scientific and government discourses to convey to the public a uniform message of scientific and political certainty concerning levels of contamination and threats to human health and the environment. For their part, concerned residents and emergent activist groups engaged with official and unofficial messaging using terms of material uncertainty, such as were examined in chapters 7 and 8. As such, I argue that the discourses that developed around the dump fire alternately made visible and registered, or obscured and attempted to displace, the material properties of the dump, as well as of the legacy of settler colonial waste in Canada's Arctic. I will show that non-expert residents (as defined by southern Canada) were able to effectively make use of scientific uncertainty about the materiality of the dump fire to draw attention to the links between the dump fire and issues of ongoing settler colonialism and social injustice.

## GOVERNING WASTE'S MANAGEMENT THROUGH SETTLER COLONIAL EXPERTISE

Throughout Iqaluit's dump fire (the one that finally caught the attention of southern Canadians), the Nunavut government released a series of public health warnings, as well as official messages intended to reassure the Iqaluit community of their safety and the health of their local environment. A roll call of experts weighed in, beginning with the Government of Nunavut's chief medical officer, the Iqaluit fire chief, and city counsellors, then expanding to include other government departments, a consultant hired from southern Canada, and administrative bodies including a freshly created Dump Fire Working Group and an Air Quality Monitoring Group. The two groups were made up of representatives and officials from relevant Government of Nunavut and City of Iqaluit departments, Aboriginal Affairs and Northern Development Canada (AANDC), the Iqaluit International Airport, Health Canada, Environment Canada, and the Department of National Defence. Neither of these assembled groups included non-experts (meaning that these groups did not include any non-southern-trained experts, for instance Inuit residents with traditional ecological knowledge).

Within a week of the spontaneous ignition of the fire, the Government of Nunavut released a series of public service announcements (PSAs). Experts such as the chief medical officer of health expressly framed the contaminants emerging from the dump fire as isolatable, calculable, and controllable. When outside, members of the public were advised to limit physical activity, and were told that they might experience symptoms including wheezing, tightness in the chest, dizziness, and/or shortness of breath (CBC News 2014b). People with asthma and/or emphysema or other breathing conditions, young people, and elderly people were advised to stay indoors, unless going outside was absolutely necessary (Varga 2014b), a situation that public health officials described as "shelter in place" (Public Safety Canada 2015, 1). According to initial reports, the health effects of the smoke were expected to be minimal as long as community members kept a safe distance from the dump (defined as approximately seventy metres; beyond this distance, contaminant concentrations were reported to be generally low, except for the odd "smoke spike" of fine particulates (Varga 2014b, unpag). These fine particulates included carbon monoxide, volatile organic compounds, hydrogen sulfide, PCBs, polycyclic aromatic hydrocarbons (PAHs), and heavy metals (Government of Nunavut 2014a; Varga 2014f).

At this time, Iqaluit lacked both an air quality monitoring system and qualified personnel who could run the required air monitoring equipment. Beginning in mid-June, air monitoring equipment was brought to Iqaluit from southern Canada to provide hourly and daily analysis of contaminants in the atmosphere.[3] These contaminants included dioxins and furans known to be carcinogenic and harmful to human development even at low levels (Health Canada 2006).[4] Data on contaminant levels was sent to southern Canada for analysis, a process that took approximately thirty days before the results were returned to Nunavut officials (Zahara 2015). During this time, Environment Canada told local residents that while the atmospheric concentrations of the most harmful contaminants (the dioxins and furans) were as yet unknown, people were in "little immediate danger" (Varga 2014f, unpag).

When the dump spontaneously caught fire in late May, Iqaluit's fire chief told the community that the fire would be left to burn out on its own, which was the best way to protect community health and the environment (Murphy 2014a). But by June 10 (less than three weeks after the dump fire began), at least ten people had gone to hospital reporting dump smoke–related health issues, and at least ten people had been hospitalized (Varga 2014f). In consultation with various experts including a consultant flown in from southern Canada, Iqaluit City Council reversed its decision, and directed the fire chief to "put out the fire in the way he sees fit" (Varga 2014g, unpag). The situation changed again on July 17, when the results of the contaminant monitoring returned from analysis in southern Canada with the finding that dioxin and furan concentrations were above Ontario's standard (Health Canada 2014, 3).[5] Contaminant analysis later revealed dioxin levels the day before had been *eleven times* Ontario's provincial limit (ibid.).

Throughout the dump fire period, government officials distinguished what they termed "facts" from "perceptions" about the health risks of smoke inhalation. One government official, for instance, remarked, "We will be able to see it [contaminant levels]. Quantify it ... These are facts, but the exposure or, basically, the [risks of] exposure to our health, are perceptions" (interview with government official, 27 August 2014). Statements such as this were based on a form of risk governance that emphasizes what is known and knowable (primarily through scientific knowledge and statistics; see chapter 2 and Zahara 2015, 2018). To better understand the context of what was going on with the changing PSAs, we need to explore the complex relationship between experts and expertise, members of the public, and trust.

## PUBLICS, EXPERTISE, AND TRUST

Research suggests people generally have a low level of scientific literacy (Miller 1991, 2004). In response, Western governments spend millions of dollars trying to increase the public's scientific literacy, using school curricula, television programs, and public lectures to convey messages about the goals and content of scientific knowledge and study (Popli 1999). Governments do this because it is widely accepted that familiarity with scientific concepts and principles is a prerequisite for effective democratic decision-making (Miller 2004). To give one example, Sturgis, Cooper, and Fife-Schaw (2005) conducted a large empirical study of peoples' knowledge of biotechnology and found that scientific knowledge has a positive effect on individual and group attitudes toward genetic science. Thus, calls are made for increased public understanding of science on the basis that members of the public (1) have little understanding of science, (2) need education about science, and (3) will more strongly support science once they have been educated (Field and Powell 2001; Wynne 1992; Irwin and Wynne 1996). This is known as the deficit model.

The deficit model is associated with a top-down approach to knowledge, according to which experts (in Western science) are assumed to have knowledge that they can translate to various audiences (politicians, policy makers, media, and the general public). The success of knowledge translation is measured by the extent to which these audiences faithfully take up this knowledge as the specialists intended. Research shows, however, that knowledge does not necessarily, or even often, circulate in this way. Members of the public mediate expert knowledge. People use multiple sources of information – from the Internet, magazines, and television shows, to information gleaned from friends and family.

Moreover, people mediate information according to their level of education, social-class privilege, cultural background, gender, and other factors. Research by Powell et al. (2007) found that emotions also play a part in mediating scientific knowledge, with worry and anger being associated with perceived uncertainty about the implications of scientific research. People also mediate scientific information while considering a particular perceived scientific uncertainty (for instance, genetically modified food) alongside other perceived scientific uncertainties (such as Creutzfeldt–Jakob – "mad cow" – disease) (Townsend, Clarke, and Travis 2004). Language also mediates how scientific knowledge is understood. For example, Lorraine Whitmarsh (2009) conducted a qualitative study on people's understanding of, and attitudes towards, climate change and global warming. Whitmarsh found that people were more likely to absolve themselves of causes, impacts, and responsibility for tackling environmental issues when they were discussed under the umbrella term "climate change" rather than "global warming." The deficit model, again, does not take this relative nature of risk into account (see also Spence and Townsend 2006). The public's trust in science is also mediated by scientific controversies, when scientists disagree with each other about the validity of a scientific claim. As such, there is not a single public; there are instead, as chapters 2 and 3 outlined, several publics.

Thus, research shows that the deficit model is not particularly effective, not least because different publics have various local expertise, and this knowledge is more highly valued within local populations. Brian Wynne (1996), for example, analyzed the clash between scientists, government officials, and farmers in the UK over the effects of radiation fallout from the Chernobyl nuclear disaster in 1986. Experts, backed up by politicians, told sheep farmers in Cumbria (in northern England) to restrict the areas in which their sheep grazed, in response to radiation fallout from Chernobyl, for a particular duration. The farmers, however, were less concerned about Chernobyl (in the Ukraine), and more concerned about Sellafield, a nuclear processing plant on their own doorstep. The farmers had local expertise about the effects of this local radiation on their land and their sheep that the experts and government ignored. Wynne found that this was not a misunderstanding: the scientists assumed themselves to be the only source of expertise (and the politicians agreed), and the farmers distrusted the scientists and politicians as a result. Moreover, it turned out that the experts were wrong, and their miscalculations cost the farmers hundreds of thousands of dollars in lost revenue (see also Collins and Evans 2002).

Recent interest in (somehow) combining traditional (Western) scientific expertise and traditional ecological knowledge (TEK) in Canada is beyond the scope of this chapter (but see Sillitoe 2006; Tuhiwai Smith 2012; Todd 2015; LaDuke 2016; Kuyek 2019). One of the critiques of these attempts is that TEK is treated as a *supplement* to Western scientific knowledge rather than superseding or being at least equal to Western science. Indeed, the progress in decolonizing science by acknowledging and working with people and their expertise beyond the academy and industry is not without issues. Gwen Ottinger (2010) conducted a case study of a white community's use of buckets to monitor air toxins, and found that the effectiveness of public involvement – as opposed to the form of participatory citizen science encouraged by experts – pivoted around issues of standards and standardized practices. Standards, Ottinger found, serve as both a "boundary-bridging function that affords bucket monitoring data a crucial measure of legitimacy among experts. But standards simultaneously serve a boundary-policing function, allowing experts to dismiss bucket data as irrelevant to the central project of air quality assessment" (ibid., 244). By adopting a scientific measure of air toxicity (the bucket method), this community group was able to claim legitimacy for their environmental concerns, but legitimacy only went so far, since the scientific experts used standardization to preclude the community from full participation (because they were not scientists, the argument went, they could not produce as accurate or valid data as the scientists themselves).

Examples like these highlight how there is a potential conflict of interest between Western expertise and decolonization. In the historical Canadian context, this conflict of interest was brought to the fore when experts (the Canadian government, missionaries, trading companies, educators, and so on) attempted to assimilate First Nations, Métis, and Inuit, with catastrophic consequences. The legacy of this violent conquest and subjugation, as I examined in section 2, survives today in, for instance, the complex relationship that Inuit have with southern Canadian government systems (which depend upon southern scientific expertise) regarding waste.

### BACK IN IQALUIT

The Canadian federal government departments charged with advising the Government of Nunavut on matters of public health employ a straightforward deficit model. Health Canada, for instance, distinguishes between "*actual* factors that affect people's level of risk" and "risk beliefs" – understandings of risk that stem from what they term "psychological" and "social" factors (2005, 5; my emphasis). And while public participation in risk management is encouraged as a form of "best practice" within Indigenous and other communities, this is done explicitly to "increase trust and understanding" of a given issue, as government officials define that issue, rather than, for instance, to openly and critically discuss and evaluate techno-scientific-bureaucratic claims about health and safety (Health Canada 2010, 1, 10).

One of the points that peppered the discussions that Alex and I had with respondents in positions of authority was the need they expressed to release consistent messages to non-expert local residents. For instance, the chief medical officer noted that then Prime Minister Stephen Harper was coming to Iqaluit as part of his annual northern publicity trip, which liaised closely with military operations in the Arctic: "Plus, Operation NANOOK's coming up, so they need to know – have the air quality monitoring data – need to know what's going on. You know, kind of blend their messages to the troops with mine."

At the one public meeting in Iqaluit to address local residents' questions and concerns about the dump fire, the chief medical officer emphasized that officials were providing consistent and regular messaging about health concerns emerging from the community:

> *the messaging does not change.* We will rebroadcast our messages. When we put out our announcements we're in English and Inuk. We do do them in other languages so people can understand. I know everyone doesn't have internet. All online through various GN routes. Also have Q and A available on our website. PSAs do go on the radio in different languages. (Public meeting, Iqaluit, 25 August 2014; Alex's field notes, my emphasis)

Note that "other languages" refers to the first language of Inuit, Inuktitut. In fact, the PSAs did change; community residents were in fact given shifting messages concerning their health and safety. As well, the government official whom I quoted earlier as distinguishing facts from perception went on to elaborate the contention that non-expert members of the public make claims such as:

> "I believe I'm affected," but questions, are you really affected? And in your lifetime, will you be affected by what has been produced due to data – what we produce? Versus what you perceive can happen to you. While the water is physical, we'll be able to see it – I hope we'll be able to see it. (Interview with government official, 27 August 2014)

Note the hesitancy of the statement, "I hope we'll be able to see it." As I will shortly demonstrate, government officials and scientists quite reasonably punctuated their declarations of scientific certainty with numerous conditional, qualifying statements that suggested uncertainty concerning both the complex and the constantly changing material conditions of the dump fire, the smoke, the surrounding water, and the contaminants in the atmosphere. As such, the messages that were explicitly meant to convey certainty, actually implicitly conveyed uncertainty – not because the (Western) experts were unsure of themselves, but because of the uncertainty of the *materiality* of the dump fire itself (and its health and environmental consequences). It was reasonable for people in positions of authority to acknowledge uncertainties surrounding health and safety issues related to the dump and the ongoing fire(s) – far more reasonable, I argue, than attempting to convey to local residents – in whose community the dump and its fire(s) burn – that their health and safety was not compromised, and then further, to criticize local residents for questioning these official statements.

Nunavut's chief medical officer placed a similar emphasis on certainty in messaging. This medical officer described the shift from no messaging at all (because this was not the first time the dump had caught fire) to realizing the need to issue a PSA, and then subsequent PSAs:

> Landfill's on fire and smoke's coming into the community – if it's going to be short term, I'm not too concerned, although as soon as there was smoking [*sic*] coming into the community, we put out an advisory. You know, that basically said "shelter in place," which is go indoors, shut your doors, all that stuff. As a way to protect your health. At that time the issue was the particulates and short-term health effects to people with chronic lung and heart disease. Then as it kind of turned chronic – the fire kept burning – then we had to look at doing a bit more. And that's when the air quality monitoring comes into place, and then, you know, working with our federal Environment and Health Canada on that monitoring – working with them on interpretation and then advising the public based on that. (Interview conducted 8 August 2014)

Defining her job as "to protect the health of the public," the health officer expressed concern that members of the public did not understand the health advisories:

So, they [the public] don't really understand that. You know what? Smoking is a health risk. I advise people not to smoke, but it's not an emergency. So it's the same thing – it's not an emergency, although there are some risks, even though the risk is very low. My job is to give the public the information I can, then they can choose whether to take that. I'm not ordering people to stay indoors – I'm saying if you want to decrease your exposure to smoke, this is the way you can do it.

Given the enormous educational campaign on the part of the Canadian government and other governments, as well as supra-state institutions such as the World Health Organization (WHO), to tell people that "smoking kills" (the WHO's more specific statement is that "smoking kills up to half of its users" (2020, unpag), local residents might reasonably be confused by the chief health officer identifying smoking as a very low risk, as well as draw an association themselves between the dump fire and cigarette smoke, and/or the gravity of the negative effects of the dump fire on human health.

Even though the medical officer thought that she presented the messaging as unambiguous, she found that things quickly became more complicated:

> because I'm saying the risk is very, very low but if you want to protect yourself, this is what you do. But people only hear the "stay out of the smoke" part. And when you say dioxins and furans, that's fear from the start ... So I find that – and I mean, *I don't go out and talk to everyone in the community* – so I go about my normal daily life. You know, *my impression* is that women of child-bearing age are not concerned, although there still are people concerned. And you know, there's pregnant women who've talked to their doctors, so I've made a point of keeping the physicians – my colleagues – informed ... But *I've also found that people don't read the Q&As. They don't read the public health advisories, they don't particularly listen* – that they hear somebody say something about how awful it is and repeat it. So, it makes it very difficult to get the messaging across. (My emphasis)

This response articulates the anecdotal-based assumption that non-experts do not avail themselves of (settler colonial–defined) expert information (that is, "they" do not read or listen). If they did, the presumption of the medical officer would be that it would lead to "people" following her expert advice, and ceasing to complain about the health effects of the dump smoke – as the deficit model assumes. At the same time, this medical officer used various qualifying terms – such as "my impression," "I don't go out and talk to everyone in the community," and "I've also found that" – that undermine the purported certainty of her own message.

Another official expressed concern that local residents did not know how to interpret messages regarding risk. Pointing to a PSA that cautioned women who were pregnant, may become pregnant, or who did not yet know they were pregnant that breathing the dump smoke might lead to decreased fertility in male offspring, she stated:

> So now people think that if they breathe in smoke that suddenly their kid's going to be sterile. No. I mean, you'd have to stand in it every day all year round. You're probably getting more dioxins from the air in Toronto than you would ever get from this fire ... So again, that's where the rationality comes in. *Like, you have more chance of getting skin cancer than you do affecting your unborn children ... I mean, in the North anyway. There's no ozone.* So anyway, that's one of the interesting aspects of the job is just the direct access to the public and no filters and no – for lack of a better word – "experts" to... spin is the wrong word – it keeps coming in my mind but it has such negative connotations. But to interpret things for public consumption. Like, not everybody's an engineer. Not everybody's a chemist. So, a chemist might talk about carcinogens in a different way than is understood by the public ... There's a lot of speculation and there's nothing wrong with that in and of itself but because you're in a public forum, it's gonna get taken the wrong way. (Interview conducted 29 July 2014; my emphasis)

Of course, neither the fire chief nor the city council could say how long the dump fire would smoulder and smoke. Nor could anyone say what materials had accumulated in the dump over the twenty-plus years since it became an informal dump. Local residents, then, many of whom spend significant amounts of time outside, quite rationally interpreted (rational thinking being the goal of the deficit model) the medical officer as saying that, indeed, local residents who were pregnant (or those who would become pregnant in the future) might well give birth to sons with reduced fertility. What is considered as rational thinking is certainly open to debate here. As well, adding that people are more at risk of cancer due to the lack of protective ozone in the Arctic (yet another environmental disadvantage that Inuit suffer from but did not cause) is hardly comforting: it is yet another hazard that local residents are largely powerless to avoid.

What the medical officer went on to say did not help to clarify the situation:

> Now, keeping in mind that if you drank one glass of water with dioxins in it, you're not going to get cancer. Like, *I think* that's a misperception – if you go out – like, what people seem to not realize is that *every time you sit around a campfire, you're breathing in enormous amounts of dioxins*. Just – it's not just plastic that produces dioxins. It's incomplete combustion of anything that produces dioxins. It's just in varying quantities. Like, chlorinated compounds produce a lot of dioxins versus just biological things or whatever. So that's another seemingly... *it's kind of* a misperception I think. A public misperception. They hear "dioxins," "cancer," "oh my god, I'm going to get cancer." You know, it's not the case. *No cancer happens overnight, you know, except radiation I guess. So, um, if you have a huge dose of radiation, you would get cancer pretty much immediately. And that would be a specific type of radiation – gamma or whatever it is.* (My emphasis)

Responses such as these might well lead to further questions, concerns, and uncertainty. The expert employs a number of qualifiers: "I think," "it's kind of." She also opens up new vistas of concern: breathing in "enormous amounts of dioxins" while sitting around campfires and getting cancer "immediately" from radiation. These new hazards may well only serve to amplify the existing risks to which local residents consider themselves vulnerable. Indeed, these statements did garner much concern from local residents. In response to local residents' concerns about the relationship between pregnancy (or potential pregnancy) and contamination, some of the respondents in official positions of authority discussed whether information concerning health risks should be screened – that is, whether only *partial* information should be released to the public. This potential screening would be justified on the basis of non-experts' supposed ignorance about how to correctly interpret scientific findings. The health officer, discussing a PSA that a colleague had issued in her absence, expressed concern that the announcement should not have included some information about the health risks of the dump smoke for pregnant women:

> A PSA was issued from the Health Department that said, "here's the results of the air sampling, but by the way, there [sic] been a few spikes here and here" – I mean, I'm paraphrasing – "Okay. All good. But, it's lower than the cancer-causing guideline but it's been above the Ontario guideline occasionally. Oh, and by the way you can get cancer from dioxins and if you're pregnant it has massive effects on whatever." And it's just – oh, hang on a second, did that other part need to be in there? I'm not so sure. Like, from my… the way I read it, that is fear-mongering. You're putting in something that, in my opinion, you know is going to be misinterpreted. That was irresponsible. But, playing devil's advocate for myself, maybe she was looking at it like, "well, I need to give all the information" … But the way it was interpreted, I can tell you, that weekend, everybody I talked to was like, "You mean to tell me that my wife has to stay inside all the time? Twenty-four hours a day?"

As I will shortly detail, local residents were aware, and concerned, that Nunavut and Ontario air quality guidelines differed (Ontario's guidelines are more stringent). This in itself is cause for concern: why are higher levels of toxins in the air we breathe acceptable in Nunavut but not Ontario? Moreover, local residents would have been correct in their suspicions that official experts may not have been providing full information, given the medical officer's statement that members of the public need not be fully informed because they may "misinterpret" information. As it was, the official on duty at the time released a PSA with more complete information, which led the medical officer to having to field questions that weekend from concerned residents.

Another city official questioned the efficacy of informing members of the public about contamination:

> And same as what if the levels were high and we said that was in the cancer-causing range? You're telling people something you

can't do anything about. It's going to happen in twenty-five years from now, there's no particular screening for it. There's nothing you can do to change it. So, it makes it very, very difficult. Now I'm not saying that you don't do it, but you have to think this through. Are you going to do more harm than good by doing it? And that's kind of a consideration with dioxins and furans, and pregnant women, too. Is it really doing more harm than good in terms of the anxiety? (Personal interview, 8 August 2014)

Again, we find an expert official suggesting that withholding knowledge from members of the public is justified because harm has already occurred that neither experts nor the affected people themselves will be able to mitigate. Not only was it reasonable for local residents to be concerned that officials were withholding complete information, the officials were doing so on the basis that they were unable to do anything about the long-term effects of this contamination on local residents. This same city official went on to suggest that the real concern about health effects should be focused on people's stress from worrying about potential contamination, rather than on their worries about the contamination itself:

Because particularly some pregnant women are very concerned. And anxiety and stress does have health effects. So, I didn't want to downplay that. So even though I think the risks are low, there's people who are very concerned, and we need to acknowledge that.

Managing populations at risk through the governing technique of messaging, then, becomes a way of shifting the discourse away from a concern with the materiality of the controversy in question, and toward how closely public perceptions of safety adhere to official government and scientific declarations of safety. When officials discuss the efficacy of selectively informing the public about health risks, the public may well wonder if what officials decide *not* to disclose may also have harmful consequences. Indeed, the public may well object to the power differential that enables experts to decide what to disclose. In the context of the devastating history of settler colonialism in the Arctic – during which Inuit were consistently lied to – Inuit communities may be particularly attuned to, and skilled in discerning, selective messaging. There is a second issue here as well:

as much as government officials and experts emphasized local residents' misinterpretation and/or emotional reaction to the presented facts – in other words, the ways in which members of the public were more focused on their own emotional perceptions than the scientific evidence – it was actually the very *materiality* of the presented (and unreported) facts that galvanized local residents to protest.

### DUMP FIRE MATERIALITY

Dumps and landfills are, nationally as well as globally, the most common form of waste disposal. While landfills, by definition, are engineered with some form of barrier between the waste and the surrounding environment, dumps are merely holes in the ground (which may be natural or the result of industry, such as an abandoned mine) or simply open spaces where waste is left. Both contain all sorts of things – diapers, dead pets, plastics, food, fabric, furniture, wrapping paper, batteries, appliances – mixing so-called hazardous with non-hazardous waste. As we know from the first section of this book, hazardous industrial waste is typically classified in a separate category than municipal solid waste (MSW) on the basis that it is generated from different sources. But as Wynne (1987) reminds us, MSW may have higher levels of toxins than industrial waste, depending on what the MSW site includes.

One of the issues of most concern is that, typically, municipalities (and even landfill operators) do not know the exact contents of any given landfill; this is even more the case with dumps. According to federal government documents, the exact origins, depth, and diversity of the waste in the West 40 site are unknown (PWC 1992; ESG 1995). However, the abandoned military equipment in some parts of the dump hint at Iqaluit's origins as an American and Canadian military base in the 1940s during Cold War fears of Soviet attack (see chapter 5), as well as concerted efforts toward sovereignty aspirations. The Nunavut government issued its "shelter in place" health advisory because, as it admitted, it did not know what was burning in the dump (Varga 2014f). Moreover, because Iqaluit is only accessible by air (and sealift during the summer), waste accumulates and is rarely removed. Iqaluit's dump, which became the city's active waste disposal site in 1995, was supposed to be a temporary solution to the city's waste issues. The site is well over capacity now.

Moreover, although scientists are familiar with subterranean fires (for instance, Australia's Burning Mountain fire has been smouldering for at least six thousand years), researchers are concerned that myriad forms of industrial waste are increasingly being placed in both dumps and landfills, and this is leading to chemical reactions that cause waste materials to combust and catch fire (Karidis 2016). These chemical reactions occur beneath the dump or landfill's surface, and little is known about them, including what causes them. As one engineer put it, "the heat-transfer processes and the chemistry that takes places [*sic*] in smoldering is significantly different from flaming fires on the surface. We know very little about smoldering. We know even less about the deep ones" (in Koerth-Baker 2016, 1). Referring to modern landfills, Maggie Koerth-Baker outlines the potentially dangerous relationship between subsurface waste fires and leachate:

> High temperatures damage the plastic liners and the pipes that keep trash-infused water called leachate from seeping into local groundwater [dumps do not have these safeguards]. High temperatures decompose trash differently and result in leachate that's harder to treat and make safe, even when it is being properly captured. And high temperatures release dangerous (and malodorous) gases that cooler trash doesn't. Those gases can find their way out of the landfill more easily because the hot spots tend to collapse, leaving cracks in the layer cake's smooth surface. (Ibid.)

Fires can smoulder at 40 to 140 feet below the surface of a dump or landfill, in wet, leachate-soaked places where fires are not normally thought to flourish. And rather than burning quickly, the slow smoulder releases a combination of combustible gases, which fuels the smoulder and releases them into the atmosphere.

Thus, the combination of a very large over-capacity dump (with unknown contents), the spontaneous combustion and burning of that dump, smoke issuing over the community, leachate running into the city's estuary which opens into the Arctic ocean, a very limited budget for tackling the fire, and a technically complex and uncertain method for either containing or extinguishing the fire (let alone managing the dump), certainly presented experts and officials with a "messaging" challenge, and local residents with grave concerns for their health and for the environment.

## MATERIAL POLITICS: MOBILIZING UNCERTAINTY

While Iqaluit's dump burned over the summer of 2014, local residents encountered first-hand what it means to live in a risk society (Beck 1992): not only did official discourses of risk assessment emphasize known risks at the expense of unknown risks, but they also shifted attention from prevention to permissible levels of contamination (Hale and Dilling 2011). It is worth recalling here Ulrich Beck's observation that to live within modernity – that is, to live with the progress that industrialization and capitalism has produced – is to live with a certain amount of contamination. As Beck notes, "The really obvious demand for non-poisoning is rejected as utopian. At the same time, a bit of poisoning being set down becomes normality. It disappears behind the acceptable values. Acceptable values make possible a permanent ration of collective standardized poisoning" (1992, 65). Recall that Nunavut's official "acceptable" level of airborne contamination is higher than that of Ontario.

As Beck (1992) theorized, contemporary risk society is characterized by members of the public who know that inherent risks accompany what politicians and industry term "technological innovation," and are increasingly wary of assessments that focus on short-terms risks rather than intermediate and long-term risks; in other words, people are skeptical when official messaging underplays the indeterminacy of risks. Indeed, the extensive risk literature notes that members of the public are increasingly being asked to deliberate on environmental, health, and other issues for which the risks are inherently indeterminate (MacFarlane 2003). As such, Michel Callon (1999) argues that members of the public make a *rational* decision *not* to trust governments who do not address the indeterminacy of risks that potentially endanger society.

Non-expert (as defined by Western ideology) Iqaluit residents were not so much a naïve or ignorant public, in need of (Western) education in order to comply with (Western) expert advice, as they were a concerned and astute assemblage of people from different backgrounds, all attempting to make sense of the messaging about health and environmental risks, the physical experience of contending with the dump smoke, and their wider experience with living on the land, as well as their intergenerational experience with ongoing settler colonialism, and all of the other information available, sourced from the local media, family and friends, the Internet, and the wider community.

During the "dumpcano" controversy, the shift by officials from addressing long-term, unknown, and perhaps unknowable risks to focusing on short-term, known, and – most importantly – permissible risks was most evident in the discussion of contaminants in the fire smoke that hovered over the community. One of the major issues that residents were concerned about was the issue of potential furan and dioxin contamination. As we saw, the official messaging, according to what the chief medical officer stated at the public meeting, was:

> *Everything except* the particulates and dioxins and furans were below public health risks ... Health Canada says in their report that they don't expect health effects in Iqaluit [and that] the way to protect yourself is the same ... Whether we found out today or in two weeks – and that's to avoid breathing in the smoke. (Alex's field notes, 25 August 2014; my emphasis).

The statement's emphasis is on those particulates (which remain unnamed) that did not presently pose a significant health risk, rather than on the two most contaminating and toxic particulates that were above the expert-defined risk threshold. While the message in the statement is meant to convey facts, certainty, and reassurance, it actually contains qualifiers – "except," "don't expect," "whether we found out today or in two weeks" – that convey a message of provisionality and uncertainty, especially concerning intermediate- and long-term risks.

Moreover, the statement raises a number of questions, such as what exactly "everything" constitutes, whether or not any particulate concentrations other than those of dioxin and furan might become elevated above the standard for health risks, and how these standards are set. This latter concern came to the fore for two reasons. First, residents were told that the standard for determining dioxin contamination was based on animal tests. The chief medical officer, in the interview, clearly identified a level of uncertainty with regard to the applicability of setting contamination thresholds for people based on animal tests: "looking at the lowest health effect ever that they can detect from exposure of pregnant rats, or I don't even know the animal species. To this then, they found a decreased fertility in the next generation of rats. That's pretty nebulous stuff." Members of the public were thus wary of scientifically described risks – which are, by definition, known risks typically involving numerical thresholds of acceptable environmental compromise – and wanted to focus on both critical evaluations of how these thresholds are determined, and unpredicted and long-term effects (Wynne 2006, 216).

Second, the different contamination standards between Nunavut and other parts of Canada also became an issue. On 16 July 2014, the dioxin concentrations were eleven times higher than the Ontario provincial limit, though this was not known until sometime later because samples had to be sent to southern Canada for analysis (Iqaluit lacks the infrastructure and on-site expertise). However, dioxin and furan sample analysis for July 17 were available, and these showed concentrations higher than the allowable amount for Ontario (an average of 0.1 picograms per cubic metre in a twenty-four-hour period). This finding triggered the Government of Nunavut's health advisory warning to pregnant (and potentially pregnant) women, described above. Not only was it unclear why the sample data analysis was available for 17 July and not 16 July, but the obvious question became why the Iqaluit community was told that the risks were low when they were, apparently, significantly higher than standards set for Canadians living in southern Canada. Indeed, there are different acceptable limits of contamination for southern and Arctic communities: in Ontario, the twenty-four ambient air quality criterion equals 0.1 picograms per cubic metre of air; for Iqaluit and other Arctic communities, it is 0.2 picograms per cubic metre of air. For at least some community members, the contrast with southern Canada reinforced their contention that a two-tier system continues to operate in Canada.

Local residents were concerned with mid- and long-term risks, and expressed this in their questions to officials and experts at the one public meeting concerning the dump fire. They asked about the prevention of future fires, and the fire chief responded:

> FIRE CHIEF: We are not separating [materials going to the dump], as is standard practice in southern landfills. We were building all [the] stuff into a huge pile of garbage. [We are] trying now to separate our easily combustible products away from our heat source. Basically, our organics from our kitchen. Basically, a while ago, human waste, trying to separate our combustibles and heat source. So, sacrificing one area to save another. We are hoping this doesn't happen.

IQALUIT RESIDENT: There's hope but what about technological solutions to detect and prevent future fires?
FIRE CHIEF: [I] believe right now, all our focus is on extinguishing the fire. I don't have an answer.

At the public meeting, members of the public raised a number of questions and concerns regarding the rationale for placing the air quality monitor (the only one available) in one particular part of the community and not anywhere else – including this exchange:

IQALUIT RESIDENT: Being that I live near the breakwater, I am quite close to that circle of safety. Is it very different – is my air quality worse than say someone up in the plateau?
DEPARTMENT OF ENVIRONMENT REPRESENTATIVE: I can't say for sure because I don't have air monitors down at the breakwater.
IQALUIT RESIDENT: I'm just trying to understand why you wouldn't have air monitors around closer to where that circle of safety is?
DEPARTMENT OF ENVIRONMENT REPRESENTATIVE: Um, we discussed with Environment Canada and Health Canada where best to site these monitors within the community. They were approved from government considerations, including accessibility to the public. We only want sites that are going to be safe and not in somebody's neighbourhood.
IQALUIT RESIDENT: So, you're saying that my area's not safe, then?

Environment and Health Canada officials determined the testing sites without consulting local residents. Related questions concerned the length of time it took for the air quality monitoring data to be analyzed in southern Canada, then for the results to be communicated to officials in Iqaluit, and then for these results to be released to the public. Far from revealing a naïve or ignorant non-expert public, the questions and comments suggested a self-informed, astute, and experienced public concerned with risks, and who asked relevant questions for which they received poor responses from (southern) experts. Consider the following exchange:

IQALUIT RESIDENT: According to the report that Health Canada released on August 4, the first measurement of dioxin and furan spike was on June 18, which is exactly thirty days before the report was released. Is this the normal time that it takes to get the results?

DEPARTMENT OF ENVIRONMENT REPRESENTATIVE: It took some time for us to work out our monitoring system with Health Canada and Environment Canada. And then we had to take the samples and ship them south for analysis. That would be the reason why.

IQALUIT RESIDENT: Yes, but why did it take thirty days? Because the sample was taken with the machine running by June 16, and by June 18 there was the first measure of 0.2 micrograms of dioxin and furan in the air ... Why thirty days? Because it's a matter of public health. So, you have to go and inform [us] of our safety. Before that was released [on] July 18, there was no mention of any toxic chemicals in that smoke in Iqaluit. There was apparently just some minor airborne particulates. There was no mention of any toxicity. And therefore, when people heard about that, they were kind of upset that it was in the air. And the fire started and broke open in the air on May 20. *So that's sixty days of people breathing air that's filled with toxicity*. And this is *according to the results that you have*. Because like one of the citizens there mentioned, residents that live close to the sea, the amount that they're breathing is likely higher than that, because this was reported in a different place. And there was no mention of where it was measured, and what was the danger of these results. It might have been more than that – than what's presented at the moment. (My emphasis)

One of the things that the "shelter in place" messaging brought to the fore was the complex interaction of southern assumptions, such as that health risk advice could be readily generalized to Indigenous communities:

IQALUIT RESIDENT: It's berry-picking season now, and everyone is out. And I'm wondering if you've kept track of the plume – of where the smoke is going from the fire. And if there are any advisories in terms of where you should or shouldn't be picking berries. Or if we should be picking berries anywhere here? What would your advice be?

DEPARTMENT OF HEALTH REPRESENTATIVE: ... I've asked Health

Canada for their advice. So, they [Health Canada], the people that do the air quality monitoring, have passed the message on to the risk assessment people and their food safety people. And their advice for me is that their pollutants aren't absorbed into the fruit, and you can wash the berries off.
IQALUIT RESIDENT: Sorry, so my question is, have you tracked the plume so that I can pick berries where my kids can eat them, where they're okay? Have you been keeping track of where the smoke has been going so that we know whether the berries are safe?
DEPARTMENT OF HEALTH REPRESENTATIVE: I think it's fair to say that the smoke has gone everywhere around this area. The wind has changed all over the place. So, my advice is to wash the berries before you consume them. And I would recommend that whether or not there was a dump fire.
IQALUIT RESIDENT: Given that we do eat berries as we pick them, is that going to be in the PSAs?
DEPARTMENT OF HEALTH REPRESENTATIVE: Yes, we can put that in the PSAs. But I don't think it's quite berry picking time yet, so we'll have time to do it.
IQALUIT RESIDENT: I was [just out] picking berries, which is why I was late.

Not only did the health officer not answer the local resident's question directly, but she appeared to lack certain knowledge – that people are out on the land picking berries (a fundamental part of traditional Inuit living), and that people commonly eat berries as they are picking them. In one of Alex Zahara's interviews with a long-time *Qallunaat* resident, the resident raised two further issues regarding Arctic community living – the fact that children (and indeed adults) spend much of their time outdoors, and the chronic and profound housing-shortage problem:

RESPONDENT: But also, okay, do they understand – are you going to tell Inuk kids to stay inside for an entire summer? What kind of... what are they going to do? Kids play outside. That's what kids should do. That's healthy. It's healthy to play outside. So, this idea that we can just go inside and shut the windows and doors – that is not a practical offering when there's seventeen people living in your house. You know? And the same thing

tonight, the statement [from the health officer] that nobody's picking berries yet. We're picking berries.
ALEX: Yeah, the berries have been out for about a week or two weeks.
RESPONDENT: Depending on where you are. And definitely people are clam digging, and definitely people are seal hunting. And we had two huge clam tides. Maybe it doesn't accumulate yet. Maybe that's not a good time to worry about it. I think with the berries, it is. That this is the time to worry about it. Is this useful information? Is the information that they're conveying usable to us in the kinds of things that we do? There's a gap.

These questions and concerns speak to the complexities and subtleties of discourses about risk in the context of the experiences of Inuit and of ongoing settler colonialism. Inuit live with the consequences of official levels of acceptable contamination that are different for Arctic and southern communities. Inuit live with southern assumptions that hunting and berry picking are lifestyle choices rather than essential parts of Inuit survival and ways of thriving (Watt-Cloutier 2015). Inuit live with levels of poverty, food insecurity, unemployment, drug and alcohol addiction, adverse health issues, and suicide that are far higher than in southern communities. As such, local residents act rationally when they question, and are skeptical of, official government messaging.

## MATERIAL POLITICS

The "dumpcano" controversy occasioned the entanglement of competing discourses centred on different degrees of certainty regarding the contamination risks to human health and the environment. The official messaging was, according to government representatives, local officials, and scientists, based on the material characteristics of the dump, the dump smoke, and the leachate, as well as scientific data (mainly derived from air quality monitoring technology, and laboratory analysis in southern Canada). Moreover, the messaging was intended to convey the clear and consistent facts known to scientists. Officials focused on local non-experts' lack of understanding and expertise, for which various solutions were considered, including deliberately withholding information. For their part, many settler

Canadians living in southern parts of Canada considered the Iqaluit dump fire to be a by-product of poor leadership on the part of the city. Years of improperly managed waste – the mixing of plastics, paper, cardboard, food waste, batteries, and even human sewage at the dump – produced what the West 40 landfill engineer emphatically described as "one of [the] worst landfills in North America" (Sperling 2014, 2).[6]

By contrast, local residents were concerned with shifting the discussion away from known risks to more complex mid- and long-term risks and to situated knowledge (Haraway 1988), and did so by emphasizing the materiality of the dump fire. In short, they engaged in material politics (Barry 2013). As such, the physical properties of the dump (its heterogeneous mix of materials, lack of modern landfill design, leachate leaking into Frobisher Bay, and so on) gained political significance as official and local discourses were variously used to register or obscure the certainty of the risks of the dump smoke to human health and the environment. Local residents also insisted on acknowledging the relationship between the dump fire and wider social justice issues, such as resource allocation (air quality monitory equipment, qualified personnel to operate the equipment and to analyze the resulting data, money to extinguish the fire properly and construct a modern landfill, and more equal contamination standards), Inuit ways of living, and issues such as the housing shortage.

That is, for those living in Iqaluit (Inuit and *Qallunaat*), the dump fire was indicative of something more – a double standard experienced by those living in Canada's Arctic. In Iqaluit, musings among residents that "this would never happen in Toronto" were common. So too were counter settler responses from those who consider living in the Arctic "to be a choice," one that is inherently unsustainable due to high government subsidies (Jay 2013).[7] These comments contrast what Inuit activist Sheila Watt-Cloutier refers to as "the right to be cold" (in her 2015 book of the same name) – that is, the right for Inuit to continue live with the land and water that have sustained them for thousands of years, and not to be forced to move or otherwise act or live in ways that are prescribed to them by settler Canadians. As such, when the territorial and federal governments refused to provide funds towards the projected $7 million needed to extinguish the dump fire, many Inuit protested. Inuit and others advocating for improved waste technology and for the Iqaluit dump to be extinguished were doing so not necessarily out of a desire to expand consumption and capitalism, but as a way of addressing long-standing issues of inequity caused by years of settler colonialism.

EPILOGUE

# Canada's Waste Future

## WASTE FLOWS DOWNSTREAM

The twenty-first century marks a threshold in which waste – as a concept, as an excess, as a phenomenon – begins to present us with an imperative – that we refigure our relations with waste within our communities, and that we come to understand waste as constituting our environments, and as something poised to become an organizing, biophysical feature of our planet. Perhaps, if we are able to look back after climate change has run its course, we will say that waste is what the world looks like in the end. A waste world. Waste at the end of the world: radioactive, biohazardous, industrial, and military waste marking the end of the gentle, hospitable Holocene and the beginning of the industrial capitalist dividend – the Anthropocene.

Throughout this book, I have returned to the upstream-downstream metaphor to distinguish between ways of defining, determining, and managing waste that may appear to be resolving waste issues but that may in reality only be providing Band-Aid solutions (and perhaps to carry the metaphor too far, Band-Aids that we will eventually have to throw away, creating further waste issues). Worse still, these solutions may exacerbate the problem because people may be convinced that these solutions are both sufficient and effective. I have argued that within the Canadian (and certainly now global) neoliberal capitalist economic and political system, waste is vigorously defined as a problem of individual and household consumption which may be effectively ameliorated through techno-scientific innovations and an individual responsibility to do such things as consume "green" products, recycle as much as possible, and safely dispose of what cannot be recycled. The first section of this book detailed how neoliberal capitalism creates and maintains this definition of waste, and how defining waste in this particular way profoundly circumscribes the parameters of the solutions to our waste issues. By far the major accomplishment of industry and government has been to deflect attention away from the profound problem of waste generation – that of the waste for which they are directly responsible – and instead focus it toward individual and household waste (which itself is largely produced by industry and then externalized to consumers). Municipal solid waste (MSW) receives by far the most attention on those few occasions when Canadians discuss waste issues. In this way, industry has effectively and efficiently, and with a moral seal of approval, externalized the cost of, and responsibility for, dealing with the bulk of Canadian waste production to consumers (Liboiron 2010).

For their part, municipal governments largely accede to how the waste management (WM) industry and the recycling industry (which are often connected) define waste, and the solutions they recommend (these solutions, of course, always involve further WM, further techno-scientific buy-in, and further individual responsibility). With these solutions there is no attempt to ameliorate the volumes of waste going to disposal or recycling (which are the same thing in many cases) because this would undermine the waste and recycling industries' profit. And we know that in Canada, as elsewhere, waste is big business – not only is haulage and disposal a billion-dollar industry, but the land used for landfills is so lucrative that in 2001, Bill Gates became the largest shareholder in Republic Services, which owns and operates landfills across the United States. As Microsoft's profits have declined steadily over the last decade, Gates's waste business has increased by over 45 per cent (Bélanger 2007). Municipal government representatives, concerned with tax rates and (re-)elections, do not tend to look further than dealing with waste through short-term disposal and diversion. They are drawn in as well by monetary incentives that only reward recycling, such as that provided by Stewardship Ontario (see chapter 3).

But for some time now we have had sufficient research that consistently shows that recycling is not a solution to our waste issues. It is indeed a "wicked problem" (Rittel and Webber 1973) in that it creates further waste issues and further environmental degradation. As chapter 3 detailed, reprocessing materials requires a great deal of energy, often using non-renewable fossil fuels that pollute the soil and atmosphere (Center for Sustainability 2012; MacBride 2012). Recycling may release hazardous wastes into the environment through by-product emissions, and/or require the use of toxic materials. It certainly *creates* waste, which then must be disposed of. For instance, as we learned in chapter 3, recycling paper requires the significant use of toxic chemicals to remove ink, and generates its own waste – sludge – that is more difficult to dispose of than paper (US Department of Energy 2006). Moreover, recycled materials are generally only suitable for one or two subsequent uses, and usually only in lesser-quality products. And without an ongoing market for many recycled materials, the "consequence is that materials thought by the public to be headed for recycling end up in landfills" (Rowe 2012, 6). To the degree that recycling waste leads to both greater waste generation (such is the case when people consume more because they believe that their post-consumption waste is effectively recycled) and people being led to believe that recycling fulfills their obligation to be good environmental citizens, recycling has significant negative political and social ramifications as well. In short, recycling is a dirty Band-Aid. This, incidentally, is my issue with so-called "zero waste" and circular economy initiatives. Insofar as these projects rely on the recycling industry, they are not accounting for the waste and environmental pollution that recycling produces, and insofar as they do not tackle the upstream issues of overproduction, overconsumption, industry hegemony, and social injustice, they remain downstream responses to our waste issues, and therefore will do nothing to stem the tide of increasing waste generation (Corvellec 2018).

I need not claim that there is a conspiracy between industry and government: a conspiracy would actually be better because it would mean that this conspiracy could be detected and rectified. On the contrary, our current WM system is simply business as usual: the neoliberal capitalist WM and recycling industry taking advantage of unsustainable overproduction and overconsumption within current mainstream sustainability rhetoric. Recycling (and minimal small-scale sporadic efforts at reducing, refusing, reusing, and refurbishing) has become *the* normalized activity of the Canadian environmental citizen. Thus, taken together, our neoliberal capitalist system, industry and government deflection, and citizens enthusiastically attempting to alleviate their individual responsibility for the burden of all of this waste through recycling, means that waste remains a stubborn and worsening problem.

### WASTE FLOWS UPSTREAM

There is some good news in all of this: families and individuals need feel far less guilty about whether or not, or how diligently, they recycle their waste. This may be somewhat upsetting for those who use their (reported) recycling behaviours as an emblem of being a "good environmental citizen." And I am *not* suggesting that individuals hedonistically overconsume and then throw everything in the trash with abandon. What I *am* arguing is that we need to start looking at waste in a different way. We need to develop ways of identifying the upstream issues that result in the production of waste and its management, and once the upstream issues have been identified, we need to address these upstream issues head-on. Thus, I devote the remainder of this book to outlining four ways in which we can begin to move to both identifying and tackling waste as an upstream issue.

*Harnessing Uncertainty*

I have argued throughout this book that the same economic and techno-scientific apparatus that facilitates the ongoing production of excess cannot be solely relied on to set the terms within which solutions might be sought. This is not only because Canada's WM operates within a neoliberal capitalist system of relentless economic growth and profit, but also, as I have argued in section 1 of this book, because techno-scientific ameliorations are inherently indeterminate. We are vulnerable to our environment (and vice versa) as latecomers to the long-established flourishing and failing of life within a constantly changing landscape. The problem with waste repositories, be they harbouring radioactive waste, MSW, mining waste, or other industrial waste – is that their containment is always limited. Successful landfill design and aftercare, in engineering terms, extends to around twenty-five to thirty years, a mere fraction of a moment in geological and bacterial time. The uncertainties of nuclear waste disposal (whether adaptive and retrievable – as with the Canadian plan – or purportedly permanent and sealed – as with Finland) leave the implementation of technical designs mired in controversy (Durant and Johnson 2009; Solomon and Andrén 2009). Moreover, nuclear energy produces radioactive waste throughout the entire fuel cycle – that is, from the mining of uranium, to reactor decommissioning (Center for Sustainability 2012; Nuclear Energy Agency 2010), to reprocessing (i.e., recycling), spent fuel is surrounded by serious questions of proliferation and safety (Lagus 2005; UNESCO International School of Science for Peace 1998). Further, both nuclear waste repositories and landfills consume enormous amounts of energy derived from fossil fuels to sort, treat, store, and transport waste (Chong and Hermreck 2010). Incinerators and EfW facilities produce extremely toxic fly ash that must be disposed of in specially designed hazardous waste landfills. And let us remember that these landfills *are* the environment: landfilled waste is not excommunicated to some other planet. *We are just moving the waste around.*

As such, WM introduces a resilient tension between determinacy and indeterminacy. This indeterminacy means the management of waste ultimately fails. It fails to be contained, fails to be predictable, fails to be calculable, fails to be a technological problem (that can be eliminated), fails to be determinate, and fails to go away. Moreover, Canadian waste has a present and future global impact whose calculability is clearly unknowable, and presents us with a responsibility that extends beyond that of simply warning the future (Johnson 2008; Durant and Johnson 2009). The lingering sense of duty toward the future – arguably the *raison d'être* for all disposal methods – has become pitted against a host of impediments, including risk, uncertainty, mistrust, and issues with siting new facilities (e.g., ShraderFrechette 2000, 2005), in what amounts to a global exercise in the politics of "moral exposure," or who lives with how much contamination for how many generations (van Wyck 2002a, 2002b, 2005, 2010, 2013a; Heimer 1985). Warning future generations about these nuclear wastes, for instance, means "saying something about a future twice as far from us as human written culture lies in the past – or roughly the entire span of time since the ice age ... [which] seems utterly impossible" (Galison 2014).

In short, waste is not a techno-scientific problem. Nor is waste a problem of resolving uncertainty. Rather, the techno-scientific uncertainty of waste should be effectively used to engage members of the public in debates about techno-scientific practices, policies, and regulation – as, for instance, Rachel Carson successfully did with her book *Silent Spring* (1962; see also Walker and Walsh 2012). Uncertainty, then, far from closing down deliberation and discussion, may be

Figure E.1 | Trash Pack toys, an ironic example of over-packaging (Amazon.com).

effectively mobilized to open discussion and decision-making to a much broader constituency of people (Revkin 2012; Wynne 2007, 108). Inspired by Carson's pathbreaking approach, which emphasizes rather than diminishes uncertainty, I argue for a heightened ethical responsibility that attends to the indeterminacy of contemporary practices of living with waste.

As such, Canadians require information about the uncertainties of established and emerging WM technologies, and opportunities to consider society's complex socioethical relations with waste in order to situate the uncertainties of waste technologies in their wider context. This wider context includes the critical junctures of waste, modes of governance, conceptualizations of risk and uncertainty, controversies, modes of public consultation, critical relationships amongst rights-holders and stakeholders, aesthetic practices, spatialities and flows, and the disjunctures between public memory and the archival record (van Wyck, personal communication). As Shiloh Krupar astutely observes, "'making visible' and openness to indeterminacy, uncertainty, ambiguity, and the like *must remain ethically attuned to the larger problematic of biopolitical governing*: the ways life and death are organized and experienced as anything but a pure opposition, that is depleting livelihoods, spectacular suffering and residual vulnerabilities" (2013, 14; my emphasis).

## Overproduction and Overconsumption

Canadians participate in a structurally supported throwaway culture that relies on constant overproduction and overconsumption. To be part of Canadian society is to participate in behaviours that minimally impact the prevailing standards of living, and that ultimately maintain or increase mass overproduction, mass circulation, and overconsumption (Shove 2003). Consider, for example, the global response of governments to the Covid-19 pandemic, which is focusing on economic recovery through increased consumption. To wit, Bryan Paterson, the mayor of Kingston, Ontario, recently sent a request to the Smith School of Business at Queen's University asking for proposals about how to get Kingstonians to consume more:

> Several months after the initial shock of the pandemic, household consumption remains at levels that are worryingly low, and that can hardly sustain a healthy local economy. Recent observations and studies indicate that consumers are very slow to reengage in consumption and pre-pandemic spending behavior. Surprisingly, this slow return to "normal" spending habits seems more common among higher income than among lower income households. This trend is worrying for a City like Kingston with a relatively important proportion of high income households ... [The challenge is] How to *nudge consumers to reengage with their previous spending habits* to support the local economy? (Email sent 24 July 2020; my emphasis)

Note that Mayor Paterson and his office is not using the Covid-19 crisis to seriously consider how Kingston might live up to its ambition to be "Canada's most sustainable city" through its adoption of Canada's Green New Deal (Steven Moore, personal correspondence), but is instead focused on the business-as-usual (literally) neoliberal capitalist response, which is to persuade people to buy more (which necessarily means waste more). Note that this was the same response that President George W. Bush had to the 9/11 attacks, in which he urged Americans to go shopping.

Canadians are by now familiar with waste problems associated with overconsumption if only through diligent efforts to recycle. Moving waste issues upstream means understanding that far more waste is produced through production than consumption. Recall, for instance, Josh Lepawsky's (2018) observation that the waste produced *from one smelter* in extracting some of the materials used in electronics is 1.8 times larger than e-waste exports *from the entire United States of America*.

Overproduction raises profound socioethical issues about our "wastemaker" society, and in particular, the effects of capitalism's refusal to identify waste as integral to production itself (Gabrys 2007, 2011; Hawkins 2011; Kollikkathara, Feng, and Stern 2009; Lynas 2012; Packard 1960). Consumption patterns influence downstream WM responses such as landfilling, recycling, and the transportation of waste. These practices will continue unabated unless structural changes occur at the upstream level; and this can only occur if Canadians critically rethink how much production is involved in consumption – that is, how much production goes into any given object that people may subsequently consume (I say "may" here because we now have reports of objects being disposed of at a mass scale even before consumption; CBC Radio *Current* 2019b; BBC News 2018) – as well as who should take responsibility for waste production. We could, for example, compel manufacturers to apply labels to all products, detailing the amount of waste produced in the manufacture of each product. Although a challenging undertaking, it would force manufacturers to find out and declare the kinds and volumes of waste produced by each product, and inform consumers as to the real environmental costs of each product they buy. A shift to a discourse of wasting less may be one way to focus on waste production resulting from the overproduction of things. This contrasts with the push for less landfilling, which focuses only on where waste ends up (Corvellec and Hultman 2012, 297).

## "The Waste They Cause Consumers to Create": Re-structuring the Relationship between Government and Industry

If we are to begin to tackle waste as an upstream issue, then communities across Canada need to take a critical look at the relationship between government (at the municipal, provincial, and federal levels) and industry. We need to start asking upstream questions: What if we placed real limitations of the transportation of waste, banning the movement of waste across not just national borders, but regional borders as well? What if we extended the responsibility for landfills to industry and chemical producers, as well as the responsibility for the toxicity of their waste? When we move waste away from ourselves, we may think that we are ridding ourselves of our waste – that we are breaking our relationship with our waste – while in fact we are doing the opposite: through our waste we are becoming connected to complex temporal, spatial, human, and inhuman relations. As Kate Parizeau and Josh Lepawsky note, waste "embodies the relationship between individuals and the collective" (2015, 23). As such, Canadians, as a citizenry, need to carefully examine the government's relationship with the extractive industry (primarily mining, oil, and gas, which also means the plastics industry), other industries (such as agriculture), and the military, who are responsible for producing the overwhelming majority, by volume and toxicity, of Canada's waste. Canadians need to demand transparency in both the volumes and kinds of waste being produced, in the name of consumer choice, sovereignty, and national security.

It follows that we need to become alert to those public education initiatives that promote individual responsibility for waste, for all of the reasons detailed in this book, and especially because these programs obscure our government's real responsibility to act on behalf of the interests of Canadians, not of industry. As Stephanie Rutherford reminds us:

> what these quick fixes elide are the ways in which environmental destruction may have less to do with the individual (and how she shops!) and more to do with sanctioned actions of governments and industrial polluters ... Rather, the responsibility for the environment is shifted onto the population, and citizens are called to take up the mantle of saving the environment in attractively simplistic ways. This allows for the management, self-surveillance

and regulation of behavior in such a way that lays claim to the kind of subjectivity that those who are environmentally conscious wish to have, and the governing of said subjectivity which does little to address the neoliberal order which contributes to environmental problems. (2007, 299)

There are indications that some levels of government are acknowledging that their relationship with industry, and with the public, needs to change. For instance, the Ontario government has recently introduced new legislation aimed at significantly extending producer responsibility. Ontario's Waste Diversion Act, 2002 has been replaced by the Resource Recovery and Circular Economy Act, 2016 and the Waste Diversion Transition Act, 2016, which together make up the Waste-Free Ontario Act, 2016. The Environmental Commissioner of Ontario's Report, "Beyond the Blue Box: Ontario's Fresh Start on Waste Diversion and the Circular Economy" (Environmental Commissioner of Ontario 2017), explicitly recognizes a number of the problems with WM and diversion that this book has detailed. For instance, in a section entitled, "What Happened to 'Reduce' and 'Reuse'?" (ibid., 22), the report states that Ontario has historically focused heavily on the lowest priority: recycling. Using the example of soda pop packaging, the report details the shift to "once-through" containers that avoided the deposit-return system (such as the one Ontario uses for beer bottles) and were easier to ship and stack. The Ministry of the Environment and Climate Change lost a short-lived and ineffective fight with industry to retain refillable bottles. This, as the report notes, in bold emphasis, "set a powerful precedent: Ontario would only make industry pay part of the cost of end-of-life management of *the wastes they cause consumers to create*, and then only for residual waste. Municipalities would be left to pay the rest" (ibid.). That is, the cost of these once-through pop and water plastic containers was externalized to consumers (Liboiron 2010; MacBride 2012). So, as Rania Ghosn and El Hadi Jazairy suggest, Canadian taxpayers may well want to ask themselves:

> why am I not compensated for collecting and sorting waste? Why do I pay taxes to have my waste collected at the curbside and delivered to a corporation who is profiting from its conversion to economic value? Most corporations pay for the extraction and refinement of raw materials, and their transportation to and

from sites of production. Why is the corporate waste industry an exception? ... The waste management system [has] the opportunity to extract value from someone else's trash, often several times over. (2015, 96, 14)

The report also details the limitations with diversion in the form of recycling (the "Blue Box is bigger in our hearts and minds than in reality"; ibid., 31), and how the relationship between Stewardship Ontario and municipalities degraded to the point of litigation (ibid., 6). Indeed, the report admits that there was a significant conflict of interest between Waste Diversion Ontario (WDO), a non-governmental organization responsible for developing, operating, and evaluating the waste diversion programs across the province, and the industry-funded organization, Stewardship Ontario, that collected fees from industry for WM and recycling. This system introduced a significant conflict of interest because these stewards were WDO's only source of funding (see chapter 3). Thus, the report is refreshingly forthright in acknowledging the structural problems inherent in the power imbalance between industry and government, the economic costs of recycling, the environmental costs of recycling, and the fact that "non-residential waste has been all but ignored: the single biggest factor in Ontario's poor waste diversion record is the lack of attention to non-residential waste" (ibid., 34). These new laws, then, provide the foundation for a WM system with less focus on individual households, and more producer responsibility for diversion, as well as more materials that are subject to extended producer responsibility. The further laws and regulations that are needed to put the Waste Free Ontario Act into place are in development, and it will take time to see whether they create a meaningful difference in the critical relationship between government, producers, the WM and recycling industries, and the Canadian public.

### Waste as a Symptom of Social Injustice

In July 2019, my son and I revisited Iqaluit, and as we were walking along the coast of Frobisher Bay, my son pointed to the smoke issuing from the West 40 dump. Five years after the "dumpcano" controversy, *plus ça change, plus c'est la même chose*. Patrick Hossay makes the uncomfortable observation that:

> We are not going to find salvation in lifestyle changes. Riding a bike to work, recycling, carrying a reusable shopping bag, or planting a tree will not make the current global system either morally or physically sustainable. (2006, 221)

A central argument directing this book is that waste is not actually the issue: waste is a *symptom* of pernicious political, economic, and social structures; it is a symptom of social injustice. This is no more clearly visible than in Indigenous communities across Canada, which continue to survive ongoing settler colonialism, as was detailed in the second section of this book. While government officials portray the waste issues in these communities as problems resulting from geographical and technological constraints (this is especially the case in the Arctic, where community remoteness and the northern climate obviate the kind of cookie-cutter adoption of WM technologies found in southern regions of Canada), communities themselves understand waste as being one among numerous acute and ongoing problems that are the outcome of brutal settler colonialism. The crises in housing, physical and mental health, food, and employment, among others, are both chronic and highly visible. These crises, like waste, are not hidden from view in Arctic and other Indigenous communities, the way they are, comparatively, in southern communities. Waste cannot be forgotten in Arctic communities for many reasons, not least because of its visibility and proximity to these communities.

WM based in a system of neoliberal capitalism depends upon a kind of forgetting (see section 1), whereas an "Aboriginal cosmopolitanism" is about remembering (Clark 2008). Nigel Clark draws our attention to a kind of remembering that proceeds from a different kind of being in the world. When a people have lived for millennia in the same place, experience is all about paying attention to variability in the landscape. "Staying in one place," Clark argues "means to travel adventurously" (ibid., 4). This is what Deborah McGregor means when she reflects that traditional environmental knowledge "implies that the knowledge is static and connected to information gained in the past. In reality, this form of knowledge is continually evolving and expanding to incorporate new information as part of adapting and responding to current challenges" (2009, 73). This is what Heraclitus famously stated, that we cannot step twice into the same river: everything, including waste, flows.

Connecting with the environment, in other words, requires remembering the experiences and sensations of others. Some Indigenous cosmology includes inhuman animals in this experience, as well as inorganic remembering – that of weather, climate, mountains and valleys, and rivers and lakes (Assembly of First Nations 1993; Atleo 2004; Cajete 2000). This sense of remembering calls an unknowable future into the present. Remembering, in this sense, is as much about the future as it is about the past. It is this mode of remembering that can be found in an account of the Australian Indigenous way of leaving waste as it is created, on the landscape, in plain view. These acts, across generations, strike Western settlers as being the antithesis of civilization, where all waste must be scrupulously moved away from people and their communities as one of civilization's key conditions (Freud 2010). For Indigenous peoples in Australia, argues Deborah Bird Rose (2003, 2004), this casting away from oneself and burying waste – this out of sight, out of mind – amounts to "self-erasure," the performance of a lie, a refusal to witness, "the equivalent of sneaking around the country." She writes:

> The remains of people's action in country tell an implicit story of knowledgeable action: these people knew where they were, they knew how to get the food that is there in the country. The country responded to their presence by providing for them, and the remains are evidence of the reciprocity between country and people. In contrast, my teachers held self-erasure to be the equivalent of sneaking around the country. Antisocial people who do not announce themselves, and use special techniques to avoid leaving tracks or traces, are up to no good. These are people who intend harm and who have something to hide. Stories about them are well known, and the evidence of their activity, in the form of illness or death, is all around one. But the actual signs of their presence are invisible; self-erasure is part of their harmful art. (2003, 62)

Recall that in Canada, we find another mode of remembering waste, through the Highway of the Atom (see van Wyck 2010), where uranium was extracted from the mine on the shores of Great Bear Lake, transported over land to Port Hope, and then to the United States for use in the Manhattan Project, where it was eventually dropped on Hiroshima and Nagasaki during the Second World

War. Not until the mid-1990s did the Sahtú Dené become aware of what had happened. "How," asks Peter van Wyck, "do we constitute in memory something that was not fully experienced to begin with?" (2012). In a remarkable act of remembering a trauma they did not experience as such, as well as a cascade of deliberations, decisions, and actions of which they were unaware, the Dené claimed responsibility for a waste they did not entirely produce, control, or even understand as waste. In 1998, a group of Dené travelled to Japan and apologized to the Japanese people for the catastrophic destruction caused by the atomic bombs, something neither the Canadian nor American governments have ever done.

Still more Indigenous communities across Canada are engaging modes of remembering waste in ways that display an understanding of responsibility in terms radically different from the contemporary neoliberal capitalist emphasis on limiting legal and financial liability. We might well learn from the Wıìlıìdeh Yellowknives Dené First Nation, whose land is polluted with the infrastructural and gold processing waste produced by the Giant Mine operations, and who now face living with 237,000 tons of arsenic trioxide. As director France Benoit details in his film *Guardians of Eternity* (2015), the Wıìlıìdeh Yellowknives Dené First Nation is both considering and preparing ways to warn future generations of people about a contamination they did not create but have been encumbered with indefinitely. The Canadian federal government's response to this toxic waste – 237,000 tons of arsenic trioxide – is to attempt to store it indefinitely. And as Krupar wryly notes, "'in perpetuity' is ultimately an unadministrable charge" (2013, 142). In other words, it is a form of governance through a "spectacle of oversight that actually minimizes the act of governing" (ibid.). The Kitchenuhmaykoosib Inninuwug (KI) of Northwestern Ontario challenged the mining company Platinex's plans for mineral exploration on traditional KI land and water, doing so on the argument that their culture is based upon an intimate relationship (involving history and identity) with land (Ariss 2017; Ariss and Cutfeet 2011). In 2008, six leaders of the KI community were imprisoned for peacefully protesting resource development on their land. The KI-6, as they became known, were released two months into their sentence. The KI community challenged the Ontario courts, successfully arguing that Platinex's plans violated the community's rights under Treaty

9. As the KI-6 repeatedly reminded government and industry officials, their obligation to protect their environment extended to all things (for instance, the fish in Big Trout Lake) and people, not just the KI community itself (Ariss 2017; Ariss and Cutfeet 2011). Similarly, the Qamani'tuaq (Baker Lake) community in the Kivalliq Region of Nunavut challenged French multinational Areva's proposal to construct a uranium mine and store radioactive waste eighty kilometres west of their community. In the struggle to resolve this controversy, community members focused on "sites of uncertainty" that involve long-term living in environment (see Metuzals and Hird 2018; Kuyek 2019).

What can we make of these Indigenous understandings of responsibility? What they have in common is a different mode of environmental ethics that, as Smith points out, "appears unbidden even before the possibility of self-reflection and deliberate self-interest emerges" (2011, xviii). This, I argue, is an environmental ethics of human vulnerability to the unpredictable volatility of geo-biological processes in deep time (see section 3). It is a vulnerability that requires unremitting visibility and remembering, and enables a form of responsibility to human and inhuman alike.

This is not to suggest that Indigenous communities in Canada are the only dumping grounds for Canada's waste. As outlined in the "Snapshot A" in the introduction, Canada exports a significant amount of waste to other countries, including countries in the globalized south such as the Philippines, Cambodia, and Malaysia. In 2019, the Canadian government issued a statement saying that it has issued no waste export permits since 2016 (Rabson 2019), but as this book has underlined, Canada's waste is being shipped overseas labelled as "recycling." The *Basel Convention on the Control of Transboundary Movements of Hazardous Wastes and Their Disposal*, signed by Canada and other countries in 1989 (and which came into force in 1992), prohibits the export of hazardous waste (although it does not cover radioactive waste) from countries in the globalized North to countries in the globalized South. In 1995, the Basel Ban Amendment was signed, which prohibits the export of any and all waste and recycling. As this book has detailed, companies have side-stepped the Convention by labelling material as "recycling," even though it is either entirely waste or contaminated by waste. To date, the Canadian federal government has not agreed to sign on to the Amendment, justifying its decision in terms of the employment that recycling offers people in the globalized South, and by arguing that Canada lacks sufficient recycling capacity (Ruff 2019). As Canadians, we need to ask whether we are prepared to continue subjecting children, women, and men in other countries to unsafe work conditions that are illegal in Canada, and for pay that is far below a living wage – simply because we do not want to deal with our own waste.

## CANADA'S WASTE FUTURE

Until recently, research on Canadian waste has largely been limited to the natural sciences and engineering, in which the focus is on technical aspects of WM engineering, and the environmental and health effects of waste disposal. However, social scientists and humanities scholars have begun to understand waste as a symptom of upstream issues of social injustice, or, as Sarah Moore remarks, "as a lens to explore environmental politics, urban history, social behavior, social movements, capitalism, modernity, risk, regulation, and governance" (2012, 781). There is a small but growing group of superb social science and humanities scholars scattered across Canada exploring waste issues. Josh Lepawsky's work (2018) on mapping the international trade and traffic of electronic waste demonstrates an increasing emphasis on the global nature of waste and its governance. Peter van Wyck's (2010) semiotic analysis of nuclear waste highlights the indeterminacy of Canada's waste future. John Sandlos and Arn Keeling's (2017) colonial history of mining in Canada demonstrates the socio-material hybridities of waste and its associated spatial considerations. Syed Harris Ali's (1999, 2002) research examines Canadian society's predilection for techno-scientific fixes and their associated consequences. Kate Parizeau and her colleagues Amy DeLorenzo and Michael von Massow analyze waste issues such as recycling and food as a complex issue of governance. Scott Lougheed's (2017) outstanding research focuses on food recalls and waste. And Max Liboiron has published vital work on plastics waste (see for example Liboiron, Tironi, and Calvillo 2018), recycling (2010, 2013), and the connection between waste and colonialism (2018; see also Virginia Maclaren's noteworthy research on this latter topic; 2010a, 2010b).

These studies are all the more vital because they do not focus on individual and household practices of disposal and diversion, as, unfortunately, the bulk of social science and humanities waste studies still do. It is very important that waste studies researchers do not inadvertently compound the industry-created and government-sponsored lie that solutions are to be found through further individual surveillance and responsibilization. There are vital areas of waste research that urgently demand singular attention: nuclear waste, military-industrial waste, and industry waste. These areas are as complex as they are interrelated – and, of course, very difficult to research because industries and militaries fiercely guard their data on the waste and contamination they produce, leaving researchers with few, if any, avenues with which to make public these titanic waste flows. The waste from Alberta's tar sands deserves several volumes of social science and humanities research to be devoted to it, as do the oil, gas, and plastics industries. Joshua Reno's (2020) timely and notable book on military waste joins Krupar's (2013) in making meaningful forays into the extensive contaminating presence of military waste. The deep connections between the oil, gas, and plastics industries merits at least one monograph (but see Center for International Environmental Law 2019; Hawkins, Potter, and Race 2015). Indeed, plastics (especially single-use) and microplastics (less than five millimetres) are gaining important attention, and require sustained research and action.

Without more research, and without media, government, and the Canadian public focusing on these upstream issues, engineers and scientists – those few people whose job it is to remember waste – will continue to be charged with the task of increasing the scope and quality of the containment of waste, and identifying the expression of waste's inevitable toxicity. Well-meaning individual Canadians will continue to diligently recycle and dispose of their waste, and continue to increase their "green" consumption in the name of decreasing waste in what Andrew Szasz (2007) argues is a sort of "inverted quarantine" whereby people with the socio-economic means attempt to insulate themselves and their loved ones from environmental hazards rather than acknowledging humanity's interdependence and devoting their energies to fighting for a safe environment for everyone. The WM industry's profits will continue to rise, as will industrial and military waste production. Our country's waste will continue to increase in both volume and toxicity, and we will continue to wonder why, despite all of our conscientious individual efforts, the wicked problem of waste keeps getting worse.

# Notes

CHAPTER ONE

1  2,500 tonnes of municipal solid waste (MSW) represents less than a day's worth of MSW generated in Metro Vancouver (Hopper 2015).
2  In 1999, Philippine officials intercepted 120 Japanese shipping containers that were similarly found to be packed with waste. In that case, the Japanese government chartered a ship to repatriate the garbage – and vowed to prosecute the company responsible (Hopper 2015).
3  For further details, see "The Giant Mine – Our Story: Impact of the Yellowknife Giant Mine on the Yellowknives Dene – A Traditional Knowledge Report."
4  Defining waste is also an exercise in complex contradiction. In Western cultures, for instance, human placentas are defined as waste (indeed, of the biohazardous kind), which allows them to be collected for scientific research. As soon as this biohazardous waste enters the placentologist's laboratory, it is an object of study, and valued resource. By contrast, some cultures define placentas as a highly symbolic material representation of kinship and spirit. This determination leads some cultures to bury placentas in the ground, but not in landfills, and with an entirely different meaning (see Scott 2012).
5  We might argue the placenta is the first waste, prior to urine and feces. In Western cultures, placentas are often considered biohazardous waste and are either incinerated or used for scientific experimentation. In non-Western cultures, placentas are highly symbolic material/spiritual entities, gifts to the earth, and so on (see Scott 2012).

## CHAPTER TWO

1 The phrase "build the beast; feed the beast" was used by an industry representative in an interview I conducted in March 2012. The representative was referring to the fact that some waste-to-energy technologies require a constant and sufficient amount of waste to remain functional. Therefore, municipalities that adopt these technologies may face having to encourage its citizens to divert less waste, cooperate with another municipality to share the technology, or import waste.
2 Waste collection fees are funded largely through property taxes, which are independent of the number of residents in a property.
3 A Consolidated Hearings Board initially turned down the proposal, but cabinet overturned that and sent it back to the board (Ontario Executive Council 1996).
4 Other consultation has addressed legislation and policies, such as the one free bag policy introduced in 2012.

## CHAPTER THREE

1 Love Canal, a community near Niagara Falls in New York State, was the subject of a twenty-one-year Superfund cleanup of toxic chemicals dumped in a landfill. This toxic waste was covered over, and through a series of property sales ended up as the site of a residential community, including schools, playgrounds and so on. See Smith (1982).
2 This target was a substantial increase from the 2004 provincial diversion rate of 28 per cent (Ministry of Environment 2010). According to the Ontario Ministry of the Environment, Ontario's diversion rate is calculated by the total quantity of waste diverted from disposal as a percentage of the total waste diverted plus disposed.
3 This estimate is based on the tonnage of waste generated in previous years (City of Kingston 2015).
4 This said, Scott Environmental Services (which was bought by Tomlinson Environmental Services in 2014) received the contract for Kingston's curbside-pickup organics program in 2008. Scott Environmental Services was a local company and did not have the bidding power of larger WM industries. However, it was the only valid response to the City's RFP and therefore did not have the competition of larger firms.
5 By-Law Number 2000-134, a By-Law to Establish Purchasing Policies and Procedures, requires city staff to obtain council's approval to award a contract when all of the conditions under section 3.4 have not been satisfied;

in this instance, section 3.4.iv of the bylaw states that three valid responses must be received. The Solid Waste Division recommended that the Waste Management of Canada bid be accepted, as it had scored 96/100 versus Tomlinson Environmental Services' score of 75/100. The Solid Waste Division stated that Waste Management's submission illustrated that "they have the experience and resources to meet the City's requirements for these services. This company has provided these services to the City in the past with satisfactory results" (City of Kingston 2015, 64). The resulting contract was accepted for a period of two years, effective January 2016, with an optional one year extension at the sole discretion of the city.

6 Moose Creek, Ontario, is located north of Cornwall, Ontario – roughly 200 kilometres away from Kingston. The landfill is owned by Laflèche Environmental.

7 According to Annie Leonard's *The Story of Stuff* (2007), 99 per cent of purchased consumer goods in the United States end up being thrown away within six months of being purchased.

8 Prior to its acquisition by Tomlinson Environmental Services, Norterra Organics – under the ownership of Scott Environmental Group – was fined in 2011 and again in 2015 for noncompliance with ministry environmental approval (Ferguson 2011). In both instances, Scott Environmental Group pleaded guilty to environmental charges ranging from accepting too much daily waste, storing more than its limit of 100 tonnes of biosolids (i.e., sewage sludge) per day, and storing biosolids outside, contrary to ministry requirements (MacAlpine 2015). In short, the company accepted biosolids that exceeded the allowable metal limits specified in the Environmental Protection Act, and, as a result, produced and sold more than a thousand tonnes of contaminated compost, which in turn contaminated several properties where it was used (MOECC 2015). Scott Environmental paid $125,000 in 2011 and $130,000 in 2015 (ibid.).

9 The industry does not necessarily need to obtain municipal contracts, per se – in that the larger corporations have plenty of options for providing services across North America. However, industries rely on municipality contracts to bolster their profits and demonstrate financial gains to their shareholders.

10 Because the Kingston waste statistics are only reported on audit years (i.e., in 2005 and again in 2013), it is difficult to determine exactly if consumption has increased due to the blue, grey, and green bin programs or just because Kingston's population has grown (or other factors that are indeterminable).

11 Payload is the carrying capacity of a vehicle, usually measured in terms of weight.

## CHAPTER FOUR

1 This chapter adopts P. Whitney Lackenbauer's designation of the Arctic as the area in Canada north of the tree line, and Inuit Nunangat, which is made up of the "land and marine areas of the Nunatsiavut, Nunavik, Nunavut, and Inuvialuit land claims settlement areas" (2011a, 71).
2 This chapter refers to "southern" and "northern" as political designations. This designation does not obviate the fact that Indigenous peoples live in southern communities, nor is it meant to obscure the fact that settler colonial legacies exist in southern Canadian communities.
3 Given the presence of middens across the Arctic, claims that Inuit produced no waste prior to colonization draw attention to the profoundly different volume and kinds of waste that colonization brought to the Arctic. As Nungak remarks, "traditionally, Inuit society was garbage-less. All our stuff was either edible by dogs, or naturally degradable" (2004, 18).
4 As William Cronon details, the Bible is replete with references to wilderness as wasteland, "places on the margins of civilization where it is all too easy to lose oneself in moral confusion and despair" (1996, 8). Wilderness is where Moses wandered with his people for forty years, nearly forsaking God and resorting to idol worship. And it is in wilderness that Jesus Christ endured forty days and was tempted by the Devil. Adam and Eve were cast out of Eden to wilderness, where they and their descendants endured pain, suffering, and hardship. Land in "its raw state," writes Cronon, "had little or nothing to offer civilized men and women" (1996, 9).
5 *Iqalummiut* is the term used locally to describe those living in Iqaluit, and includes both Inuit and *Qallunaat* (non-Inuit) community members. *Qallunaat* is an Inuktitut word for people who are not Inuit – typically white people. See *Qallunaat! Why White People Are Funny*, National Film Board of Canada, 2015.
6 See QIA (2013a) for detailed community histories, including explanations of why colonization in Canada's North did not occur until the twentieth century. Tester (2010a) provides further discussion of socio-economic changes during this time.
7 "Iqaluit" is Inuktitut for "place of many fish." The Hudson's Bay Company established an outpost in the area in 1914.
8 Frobisher Bay was the name given to the community by European settlers until city council voted to change it to Iqaluit in 1987.
9 During this time, it was common to use the terms "native," "Métis," or "Eskimo" when referring to Inuit people. These terms are now considered

pejorative or simply incorrect. Eskimo, for example, is a Cree term meaning "eater of raw meat."
10 The term "Eskimo" refers to Inuit as "raw-meat eaters." It is a derogatory term that persists today, and is one of many signifiers of ongoing settler colonialism in Canada. To wit, the Edmonton Eskimos football team has changed its name in response to the 2020 Black Lives Matter social movement. See https://www.esks.com.
11 So stated by Alvin Hamilton, the minister of Northern Affairs and National Resources under Prime Minister John Diefenbaker.
12 This later became the Department of Indian Affairs and Northern Development in 1985, and Aboriginal Affairs and Northern Development Canada (AANDC) in 2011. Note the reiterated association between Indigenous peoples and development.
13 John S. Milloy (1999) provides a detailed history and description of Canada's residential school system from 1879 to 1986. The closing events for Canada's Truth and Reconciliation Commission on residential schools were held in Ottawa in 2015.
14 Nunavut means "our land" in Inuktitut. Prior to 1 April 1999, Nunavut was part of the Northwest Territories (NWT).
15 Prior to colonization, Inuit did not have surnames, and because Inuktitut was an oral language, first names did not have consistent spelling. In 1941, all Inuit were given E-numbers (Eskimo-numbers), which allowed the government to accurately collect "census information, trade accounts, medical records and police records" (Bonesteel 2006, 38). The E-number system ended after 1968, when the federal government, through its Project Surname, requested all Inuit select a surname to be used by government officials (ibid.).
16 The term *Inuit* means "real human beings." *Inuk* is singular.
17 As John Ralston Saul observes, "the European tradition is that you can own land, while you merely pass through water" (2008, 301). As opposed to the European distinction between land (useful, owned, sovereign) and water (for passage), within Inuit traditions, land, ice, and water are one. Sheila Watt-Cloutier, former chair of the international Inuit Circumpolar Council, pointed out: "As Canadians seek to assert our sovereignty in the Arctic, we must remember that history is on our side, that Inuit traveled an *icy highway* through the Northwest Passage long before more recent arrivals even considered a fast route west" (in Saul 2017, n.p.; emphasis added).
18 When Britain transferred what was then known as the Arctic Islands to Canada in 1880, Danish and American explorers had already begun to

circumnavigate parts of the Arctic, sometimes encountering Inuit peoples. Around this time, Greenlanders were regularly hunting muskox on Ellesmere Island, and parallels between the Indigenous cultures of both countries were apparent. The US, it was feared, would also soon lay claim to the Arctic. With this in mind, the Canadian government made a number of declarations of sovereignty, and dispatched RCMP personnel to Ellesmere Island, Pond Inlet, Baffin Island, and other regions considered most likely to incur sovereignty claims from Greenland, Norway, the Soviet Union, or the United States. This was not altogether paranoia: in 1904, Otto Sverdrup did lay claim to a region west of Ellesmere Island in the name of the Norwegian King Oscar II – a claim only settled in 1930 when Norway officially recognized Canadian ownership.

19 This presumption was necessary in order to make the very claim of sovereignty. In other words, Inuit *had* to be Canadian in order for Canada to assert its sovereignty over the land in which Inuit lived and had lived for many generations before the federation of Canada.

20 Lester Pearson, then secretary of state for external affairs, urged the Cabinet to employ some means to preserve Canadian sovereignty over the vast wastelands of the Arctic. As was stated in the 1933 International Court of Justice case, human activity was essential to sovereignty, and relocating Inuit from northern Quebec to the high Arctic was a solution to the "Eskimo problem" (Pigott 2011, 226–7).

21 The Supreme Court of Canada declared: "there was from the outset never any doubt that sovereignty and legislative power, and indeed the underlying title, to such lands vested in the Crown" (*R. v. Sparrow*, S.C.R., in Asch 2007, 283).

22 The anthem goes: "O Canada, our home and native land, true patriot love in all of us command. With glowing hearts we see thee rise, the true north strong and free."

23 Canada vies for its share of this pie with Russia, Norway, Sweden, the USA, Greenland, Iceland, and Finland.

24 See Emilie Cameron's (2012) discussion of how Inuit knowledge has been considered in social science climate change research.

25 Information obtained from Aboriginal and Northern Affairs Canada completed Access to Information Request: A-2013-01167.

26 https://www.reddit.com/r/worldpowers/comments/6xhud9/conflict_operation_nanook_2018.

27 Interview with Inuit Iqaluit resident, conducted 25 June 2014.

28 Interview with Inuit Iqaluit resident, conducted on 22 June 2014.

29 These assertions of the absence of waste prior to colonization may be

interpreted as reiterating the common colonial "noble savage" trope of the mythical Inuit living in harmony with nature – what postcolonial scholars have noted is invariably a construction of the colonial gaze (see Rose 2003 for a detailed discussion of how the "noble savage" and "dismal savage" tropes have been used in academic waste literature to silence Indigenous voices and lived experience).

30 The "Statement on Canada's Arctic Foreign Policy: Exercising Sovereignty and Promoting Canada's Northern Strategy Abroad" from Canada's Department of Foreign Affairs and International Trade (2010) devotes less space to waste concerns, but does state that "Canada will continue to address the problems arising from these contaminants [persistent organic pollutants (POPs)], including *waste management practices* in the North, and will engage actively in global negotiations to reduce mercury emissions" (in Griffiths, Huebert, and Lackenbauer 2011, 269; emphasis added).

31 Lackenbauer and Farish are citing Hooks and Smith, "The Treadmill of Destruction" (2004).

## CHAPTER FIVE

1 The title of this chapter derives from the term used by white explorers to describe land in Canada's Arctic (see Amagoalik 2000). The term "waste" is sometimes still used today, nearly always in discussions of Arctic sovereignty (see, for example, Harris 2005).
2 While it is locally referred to as a landfill, it is in fact a dump, because it lacks the engineering features of a landfill (see chapter 1).
3 For a more in-depth discussion, see Zahara (2015).
4 The dump fire was eventually extinguished on 16 September 2014.
5 Interview with *Qallunaat* Iqaluit resident, conducted on 9 August 2014.
6 In response to Environmental Petition No. 50 under Section 22 of the Auditor General Act, Environment Canada stated that "the sea disposal of military waste has occurred in many areas of the world" (Kehoe 2002, Q3).
7 By 1988, PCBs had shown up in the breast milk of Inuit women living in Baffin Island, the Northwest Territories, and northern Quebec. These PCB levels were five times higher than in women's breast milk in any southern parts of Canada, and are the highest concentration ever found in women except for those directly involved in industrial accidents (Myers 2001).
8 The Pinetree Line refers to a series of radar stations across the fiftieth parallel of the United States and Canada.

## CHAPTER SIX

1. This chapter employs the term "inhuman" to refer to living and nonliving creatures that are not included in the species *Homo sapiens*, and to emphasize that the classification itself is an evolutionary creation of an unfathomable diversity and population of microorganisms that literally make up "the human." Furthermore, doing so is more in line with Inuit and other Indigenous cosmologies, which readily challenge the human/inhuman binary. For detailed discussion see Hird (2009, 2010a, 2010b, 2010c, and 2012c).
2. This is not to say that Inuit believe they do not impact nature; indeed, even talking about polar bears is considered to affect polar bear migration patterns (see Henri 2012, 190–9). What I mean is that nature was always understood as having the ability to act outside of Inuit control. If Inuit were gone, nature would simply act differently.
3. While non-Inuit may differentiate between subsistence hunting and trophy hunting, for many Inuit the relationship is more complex. As well as engaging in regulated subsistence hunting, some Inuit engage in southern Canadian-initiated – and foreign-initiated – trophy hunting because it provides vital employment for Inuit, and the meat is given to the community. Unemployment and food insecurity are profound issues in the Arctic. Inuit repeatedly point out that they rely on country food in order to feed their families because southern Canadian food is both prohibitively expensive and of little nutritional value.
4. See also Hugh Brody's (2000) detailed examination of the differences between hunter-gatherer and agricultural ways of living, the ascendancy of the latter, and its negative consequences for the survival of hunter-gatherer societies.
5. Geoengineering typically refers to the "big science and technology" harnessed to mitigate climate change. I use the term here to call attention to the vast and complex infrastructure that are both geological and engineered, and attempt to deal with ever-increasing amounts and toxicity of waste.
6. Interview with Inuit resident of Iqaluit, conducted on 25 June 2014.
7. Using the term "commodity" recognizes that the value given to sled dogs by settler colonial government officials was largely economic. Sled dogs had economic value: they allowed Inuit to go out on the land to obtain food, clothes, or pelts for trading. Yet, for Inuit, as Tester explains, "dogs were a social entitlement with limited exchange and no market value …

their loss and replacement by snowmobiles has complex implications for Inuit culture" (Tester 2010a, 130).
8 It is important to note that a view of dogs as either wholly domesticated or feral invokes a natural/cultural distinction that is not recognized through Inuit cosmology.
9 Allowing *qimmiit* to roam free was essential for establishing pack positions and for the development of hunting skills (McHugh 2013, 164–5).
10 See, for example, the conversation between anthropologist Toshio Yatsushiro and Jamesie, an Inuk man from Nunavut in Lisa Stevenson, "The Psychic Life of Biopolitics," 604–5.
11 *Qimmiit* were most often found "foraging around a dump or breaking into storage areas" and were actually more of a danger to public property than to people (despite their sometimes aggressive behaviour) (QIA 2015, 39).
12 Indeed, the federal government has used the idea of Arctic sovereignty to justify economic expansion throughout the North. In his speech from the throne, Prime Minister Stephen Harper stated, "the eyes of the world increasingly look enviously to our North. Our Government will not rest." This was followed by promises of increased offshore patrol ships, the construction of a new highway through the Northwest Territories, and the construction of a northern scientific research station (Harper 2013).
13 De Vos provides a similar description of husky dog team areas in Greenlandic Inuit communities.
14 Interview with a long-time *Qallunaat* resident of Iqaluit, 10 July 2014.

CHAPTER SEVEN

1 The credit for this title goes to Laurence Rocher and Romain Garcier.

CHAPTER EIGHT

1 The effects of arsenic depend on the dose. Arsenic is lethal to humans at a range of 70 to 180 milligrams. Doses below this threshold produce various effects including vomiting, diarrhea, muscle pain, skin rashes, paresthesia, and keratosis. Lower doses over longer periods of time (over years) produce black spots on the skin, and indications are that in very low doses arsenic is a carcinogen, producing lung, liver, and bladder cancers. Arsenic trioxide resembles dust and is soluble in water; therefore, it is a health threat to any organism drinking from or potentially living near streams, lakes, puddles, or snow contaminated by arsenic dust. There have been a

long series of complaints, an unconfirmed number of deaths of children, illness and death among livestock and wildlife, investigations, commissioned reports and other reports, denials of responsibility, further reports, media exposés, and now a lengthy assessment process to determine what to do with Giant Mine's mammoth waste.
2 Industrialism, capitalist economies, and neoliberal governance mean that this great expunging of our waste is now global in scale and topography (Spaargaren, Mol, and Buttel 2006; Dauvergne 2010).
3 Bruno Latour defines black-boxing as "the way scientific and technical work is made invisible by its own success. When a machine runs efficiently, when a matter of fact is settled, one need focus only on its inputs and outputs and not on its internal complexity. Thus, paradoxically, the more science and technology succeed, the more opaque and obscure they become." See Bruno Latour (1999, 304).
4 Likewise, our knowledge of the effects of nuclear fallout and low-level radiation exceed what we have been able to observe and extrapolate, both spatially and temporally. For example, consider the Russian thistle that absorbs strontium-90 and cesium, and metabolizes it from nuclear-contaminated areas, its head eventually severing from its stem, thus becoming a source of windblown-distributed radiation that is difficult to track (Stang 1998).

CHAPTER NINE

1 The previous fires at the West 40 dump (December 2013, January 2014, and March 2014) were more easily extinguished because of their confinement to a small portion of the dump area.
2 *Qallunaat* (also written "*Qablunaat*"), as chapter 4 notes, is an Inuktitut word for non-Inuit people. For more details, see Cameron (2015, 22–4).
3 Who first approached whom is the subject of some dispute. Nunavut's chief medical officer at the time told Alex in an interview that she contacted Health Canada and Environment Canada, and together they assembled and implemented a plan to bring monitoring equipment to Iqaluit, and to have the data analyzed at a laboratory in southern Canada. According to a journalist whom Alex interviewed, City of Iqaluit officials approached neither Environment Canada nor Health Canada to bring the necessary equipment and test for smoke contaminants. According to this journalist, Environment Canada officials approached Leona Aglukkaq, Iqaluit's member of parliament at the time, suggesting monitoring equipment be sent and equipment training provided.

4 According to Health Canada (2006, 1), the "tolerable" level of contaminants – meaning "no serious health effects are expected" – is seventy picograms per kilogram of body weight per month, or one-trillionth of a gram.
5 This standard – adopted by the Government of Nunavut – is a twenty-four-hour average of 0.1 picograms per cubic metre.
6 Other engineers would emphatically define the West 40 site as a dump, not a landfill.
7 See also the many responses given to the *National Post* when readers were asked how to go about "solving Canada's native issue" (Russell 2013).

# References

Aboriginal Affairs and Northern Development Canada (AANDC). 2013a. "Giant Mine Remediation Project." Ottawa, ON: Aboriginal Affairs and Northern Development Canada. http://www.aadnc-aandc.gc.ca/eng/1100100027364/1100100027365.
– 2013b. "What's Happening at Giant?" Ottawa, ON: Aboriginal Affairs and Northern Development Canada. http://www.aadnc-aandc.gc.ca/eng/1374777790923/1374777851043.
– 2013c. *Northern Contaminated Sites Program Performance Report: 2010–2011*. Ottawa, ON: Aboriginal Affairs and Northern Development Canada.
Adeola, Francis. 2012. *Industrial Disasters, Toxic Waste, and Community Impact: Health Effects and Environmental Justice Struggles Around the Globe*. Lanham, MD: Lexington Books.
Agamben, Giorgio. 2002. *State of Exception*. Translated by Kevin Attell. Chicago, IL: University of Chicago Press.
– 2003. *The Open: Man and Animal*. Translated by Kevin Attell. Stanford, CA: Stanford University Press.
Agar, Betsy J., Brian W. Baetz, and Bruce G. Wilson. 2012. "Fuel Consumption, Emissions Estimation, and Emissions Cost Estimates Using Global Positioning Data" *Journal of Air and Waste Management Association* 57: 348–54.
Ahuja, Neel. 2015. "Intimate Atmospheres: Queer Theory in a Time of Extinctions." *GLQ: A Journal of Lesbian and Gay Studies* 21, no. 2–3: 365–85.
Ajzen, Icek. 1991. "The Theory of Planned Behavior." *Organizational Behavior and Human Decision Processes* 50, no. 2: 179–211.

Alexander, Catherine, and Andrew Sanchez, eds. 2019. *Indeterminacy: Waste, Value, and the Imagination.* New York: Berghahn Books.
Ali, Syed Harris. 1999. "The Search for a Landfill Site in the Risk Society." *Canadian Review of Sociology* 36, no. 1: 1–19.
– 2002. "Disaster and the Political Economy of Recycling: Toxic Fire in an Industrial City." *Social Problems* 49, no. 2: 129–49.
Allen, Barbara L. 2007. "Environmental Justice and Expert Knowledge in the Wake of a Disaster." *Social Studies of Science* 37: 103–10.
Alunik, Ishmael, Eddie D. Kolausok, and David Morrison, eds. 2003. *Across Time and Tundra: The Inuvialuit of the Western Arctic.* Vancouver, BC: Raincoast Books.
Amagoalik, John. 2000. "Wasteland of Nobodies." In *Nunavut: Inuit Regain Control of Their Lands and Lives*, edited by Jens Dahl, Jack Hicks, and Peter Jull, 138–9. IWGIA Document no. 102. Copenhagen: International Work Group for Indigenous Affairs.
– 2015. "My Little Corner of Canada, Jan. 9: Working for the GN." *Nunatsiaq News*, 9 January 2015.
Amegah, Adeladza K., Jouni J. K. Jaakkola, Reginald Quansah, Gameli K. Norgbe, and Mawuli Dzodzomenyo. 2012. "Cooking Fuel Choices and Garbage Burning Practices as Determinants of Birth Weight: A Cross-Sectional Study in Accra, Ghana." *Environmental Health* 11: 78.
Alma Mater Society (AMS). 2012. "August 14th Remarks to City Council Re: 'One Garbage Bag' Policy." Alma Mater Society of Queen's University. Accessed 20 September 2012.
Amegah, A. Kofi, and Jouni Jaakkola. 2016. "Street Vending and Waste Picking in Developing Countries: A Long-Standing Hazardous Occupational Activity of the Urban Poor." *International Journal and Occupational and Environmental Health* 22, no. 3: 1–6.
Anderson, Kay. 2000. "'The Beast Within': Race, Humanity, and Animality." *Environment and Planning D: Society and Space* 18: 301–20.
Anderson, Warwick. 1995. "Excremental Colonialism: Public Health and the Poetics of Pollution." *Critical Inquiry* 21, no. 3: 640–69.
– 2010. "Crap on the Map, or Postcolonial Waste." *Postcolonial Studies* 13, no. 2: 169–78.
Anderson, Erik, and Sarah Bonesteel. 2008. *Canada's Relationship with Inuit: A History of Policy and Program Development.* Ottawa, ON: Minister of Indian Affairs and Northern Development and Federal Interlocutor.
Arendt, Hannah. 1958. "Irreversibility and the Power to Forgive." In *The Human Condition*, 236–43. Chicago, IL: University of Chicago Press.

Ariss, Rachel. 2017. "Platinex V Kitchenuhmaykoosib Inninuwug: Extraction and the Role of Law in KI's Struggle for Self-determination." *Contours* 7.

Ariss, Rachel, and John Cutfeet. 2011. "Kitchenuhmaykoosib Inninuwug First Nation: Mining, Consultation, Reconciliation and Law" *Indigenous Law Journal* 10, no. 1: 1–37.

Arktis Solutions. 2010. *Development of an Overview of the State of Waste Management in Canada's Territories*. Ottawa, ON: Environment Canada, Waste Reduction and Management Division.

– 2011. *Report on Current State of Solid Waste Management and Facilities in Nunavut and Cost-Benefit Analysis of Selected Solid Waste Management Approaches*. Iqaluit, NU: Government of Nunavut, Community and Government Services.

Armstrong, David. 1983. *The Political Anatomy of the Body*. Cambridge, MA: Cambridge University Press.

Asch, Michael. 2007. "Governmentality, State Culture and Indigenous Rights." *Anthropologica* 49, no. 2: 281–4.

Assembly of First Nations (AFN). 1993. "'Environment' in Assembly of First Nations." In *Reclaiming Our Nationhood; Strengthening Our Heritage: Report to the Royal Commission on Aboriginal Peoples*, 39–50. Ottawa, ON: Assembly of First Nations.

Atasu, Atalay, and Ravi Subramanian. 2012. "Extended Producer Responsibility for E-Waste: Individual or Collective Producer Responsibility?" *Production and Operations Management* 21, no. 6: 1,042–59.

Atleo, R. 2004. *Tsawalk: A Nuu-chah-nulth Worldview*. Vancouver, BC: University of British Columbia Press.

Aupilaarjuk, Mariano, Marie Tulimaaq, Akisu Joamie, Emile Imaruittuq, and Lucassie Nutaraaluk. 1999. *Interviewing Inuit Elders Volume 2: Perspectives on Law*, edited by Jarich Oosten, Frédéric Laugrand, and Wim Rasing. Iqaluit, NU: Nunavut Arctic College.

Baarschers, William H. 1996. *Eco-Facts and Eco-Fictions: Understanding the Environmental Debate*. London and New York: Routledge.

BacTech. 2013. "BacTech Hires Tetra Tech for Snow Lake, Manitoba Project." BacTech. https://www.globenewswire.com/news-release/2013/01/17/1345643/0/en/BacTech-Hires-Tetra-Tech-for-Snow-Lake-Manitoba-Project.html (accessed 4 September 2020).

Ballingall, Alex. 2014. "Whitby Plastic Recycler Denies Shipping Trash to Philippines," *The Star*, 13 February.

Barad, Karen. 2003. "Posthumanist Performativity: Toward an

Understanding of How Matter Comes to Matter." *Signs: Journal of Women in Culture and Society* 28, no. 3: 801–31.

– 2007. *Meeting the Universe Halfway: Quantum Physics and the Entanglement of Matter and Meaning*. London, UK: Duke University Press.

– 2011. "Erasers and Erasures: Pinch's Unfortunate 'Uncertainty Principle.'" *Social Studies of Science* 41, no. 3: 443–54.

– 2012a. "On Touching: The Inhuman That Therefore I Am." *Differences* 23, no. 3: 1–21.

– 2012b. "What Is the Measure of Nothingness: Infinity, Virtuality, Justice." *dOCUMENTA: 100 Notes—100 Thoughts*, no. 13: 4–17.

Barr, Stewart, and Andrew W. Gilg. 2006. "Sustainable Lifestyles: Framing Environmental Action in and Around the Home." *Geoforum* 37, no. 6: 906–20.

Barr, Stewart, Andrew W. Gilg, and Nicholas J. Ford. 2001. "A Conceptual Framework for Understanding and Analysing Attitudes Towards Household-Waste Management." *Environment and Planning A* 33, no. 11: 2,025–48.

Barry, Andrew. 2013. *Material Politics: Disputes Along the Pipeline*. Oxford, UK: Wiley-Blackwell and Sons.

Baudrillard, Jean. 1998. *The Consumer Society: Myths and Structures*. London, UK: Sage Publications.

Bauman, Zygmunt. 2001. "Excess: An Obituary." *Parralax* 7, no. 1: 85–91.

Baviskar, Amita. 2003. "For a Cultural Politics of Natural Resources." *Economic and Political Weekly* 38, no. 48: 5,051–5.

Baxter, Jamie W., John D. Eyles, and Susan J. Elliott. 1999. "From Siting Principles to Siting Practices: A Case Study of Discord among Trust, Equity and Community Participation." *Journal of Environmental Planning and Management* 42, no. 4: 501–25.

British Broadcasting Corporation News (BBC News). 2018. "Burberry Burns Bags, Clothes and Perfume Worth Millions." British Broadcasting Corporation, 19 July.

Beacon Environmental. 2008. *Wildlife Hazard Assessment and Integrated Wildlife Management Plan for the City of Yellowknife Solid Waste Facility*. Project 206086. Prepared for City of Yellowknife. Markham, ON: Beacon Environmental.

Beck, Ulrich. 1992. *Risk Society: Towards a New Modernity*. London, UK: Sage Publications.

– 1995. *Ecological Politics in the Age of Risk*. Translated by Amos Weisz. Cambridge, MA: Polity Press.

– 2007. *World at Risk*. Cambridge, MA: Polity Press.

Bélanger, Pierre. 2007. "Airspace: The Geopolitics of Landfilling in Michigan." In *Alphabet City: Trash*, no. 11, edited by John Knechtel, 132–5. Cambridge, MA: MIT Press.

Belcourt, Billy-Ray. 2015. "Animal Bodies, Colonial Subjects: (Re)Locating Animality in Decolonial Thought." *Societies* 5, no. 1: 1–11.

Benazon, Netta. 1995. "Soil Remediation: A Practical Overview of Canadian Cleanup Strategies and Commercially Available Technology." *Hazardous Materials Management* 7, no. 5: 10–26.

Benford, Gregory. 1999. *Deep Time: How Humanity Communicates Across Millennia*. New York: Bard.

Benford, Gregory, Craig W. Kirkwood, Harry Otway, and Martin J. Pasqualetti. 1991. "Ten Thousand Years of Solitude? On Inadvertent Intrusion into the Waste Isolation Pilot Project Repository." Los Alamos, NM: United States Department of Energy.

Bennett, Joseph R., Justine D. Shaw, Aleks Terauds, John P. Smol, Rien Aerts, Dana M. Bergstrom, Jules M. Blais, et al. 2015. "Polar Lessons Learned: Long-Term Management Based on Shared Threats in Arctic and Antarctic Environments." *Frontiers of Ecological Environment* 13, no. 6: 316–24.

Benoit, France, director. 2015. *Guardians of Eternity*. 2015.

Bergbäck, Bo, and Ulrik Lohm. 1997. "Metals in Society." In *The Global Environment: Science, Technology and Management*, edited by D. Brune, D.V. Chapman, M.D. Gwynne, and J.M. Pacyna, 276–89. Oslo: Scandinavian Science Publisher.

Berglund, Christer, and Patrik Söderholm. 2003. "An Economic Analysis of Global Waste Paper Recovery and Utilization" *Environmental and Resource Economics* 26: 429–56.

Bernauer, Warren. 2011. "Uranium Mining, Primitive Accumulation and Resistance in Baker Lake, Nunavut: Recent Changes in Community Perspectives." MA thesis, University of Winnipeg.

– 2012. "Uranium Controversy in Baker Lake." *Canadian Dimension*, 3 February.

Beyer, J., H. Trannum, T. Bakke, P. Hodson, and T. Collier. 2016. "Environmental Effects of the Deepwater Horizon Spill: A Review." *Marine Pollution Bulletin* 110: 28–51.

Bhagavad Gita. 1944. Translated by Vivekananda-Isherwood. Chapter 11, verse 32. Hollywood, CA: Marcel Rodd.

Biello, David. 2015. "How Microbes Helped Clean BP's Oil Spill." *Scientific American*, 28 April.

Bird Rose, Deborah. 2003. "Decolonizing the Discourse of Environmental

Knowledge in Settler Societies." In *Culture and Waste: The Creation and Destruction of Value*, edited by Gay Hawkins and Stephen Muecke, 53–72. London, UK: Rowman and Littlefield Publishers.
– 2004. *Reports from A Wild Country: Ethics of Decolonization*. Sydney: University of New South Wales Press.
Blake, Dale. 2001. *Inuit Life Writings and Oral Traditions Inuit Myths*. St John's, NL: Educational Resource Development Co-operative.
Blaser, Mario. 2009. "The Threat of the Yrmo: The Political Ontology of a Sustainable Hunting Program." *American Anthropologist* 111, no. 1: 10–20.
Bocking, Stephen. 2007. "Science and Spaces in the Northern Environment." *Environmental History* 12, no. 4: 867–94.
– 2009. "Defining Effective Science for Canadian Environmental Policy Leadership." In *Canadian Environmental Policy and Politics: Prospects for Leadership and Innovation*, edited by Debora L. VanNijnatten and Robert Boardman, 64–76. Oxford, UK: Oxford University Press.
Bonesteel, Sarah. 2006. *Canada's Relationship with Inuit: A History of Policy and Program Development*, edited by Erik Anderson. Ottawa, ON: Indian and Northern Affairs Canada.
Bordeleau, Erik. 2011. "The Care of the Possible: Isabelle Stengers interviewed by Erik Bordeleau." Translated by Kelly Ladd. *Scapegoat* 1: 12–27.
Bowman, John. 2014. "#Sealfie: Northerners Respond to Ellen's Seal Hunt Views with Fur Photos." CBC News, 28 March.
Boynton, Sean. 2019. "Ship Carrying 69 Containers of Garbage Arrives in Vancouver After Journey from Philippines." Global News, 29 June.
Braun, Kathrin, and Susanne Schultz. 2010. "'A Certain Amount of Engineering Involved': Constructing the Public in Participatory Governance Arrangements." *Public Understanding of Science* 19, no. 4: 403–19.
Braune, B., D. Muir, B. DeMarch, M. Gamberg, K. Poole, R. Currie, M. Dodd, et al. 1999. "Spatial and Temporal Trends of Contaminants in Canadian Arctic Freshwater Terrestrial Ecosystems: A Review." *Science and the Total Environment* 230, no. 1–3: 145–207.
Bravo, Michael. 2006. "Science for the People: Northern Field Stations and Governmentality." *British Journal of Canadian Studies* 19, no. 2: 221–46.
Briggs, Jean. 1999. *Inuit Morality Play*. New Haven, CT: Yale University Press.
Brody, Hugh. 2000. *The Other Side of Eden: Hunters, Farmers, and the*

*Shaping of the World*. Vancouver, BC: Douglas and McIntyre Publishing Group.

Browne, Heather. 2008. "Curbing my Appetite for Plastic." *Kingston Whig-Standard*, 20 September.

Bulkeley, Harriet, Matt Watson, and Ray Hudson. 2007. "Modes of Governing Municipal Waste." *Environment and Planning A* 39, no. 11: 2,733–53.

Bulkeley, Harriet, Matt Watson, Ray Hudson, and Paul Weaver. 2005. "Governing Municipal Waste: Towards a New Analytical Framework." *Journal of Environmental Policy and Planning* 7, no. 1: 1–23.

Burchell, Graham, Colin Gordon, and Peter Miller. 1991. *The Foucault Effect: Studies in Governmentality*. Chicago, IL: University of Chicago Press.

Busch, Laura. 2013. "Accounting for Trash: A Look at What and how Yellowknifers Recycle." Northern News Service, 19 April.

Butler, Catherine. 2010. "Morality and Climate Change: Is Leaving Your TV on Standby a Risky Behaviour?" *Environmental Values* 19, no. 2: 169–92.

Buttigieg, Bryan J. 2010. "Consulting Engineers Help Develop New Ontario Soil Background Data for Use in Brownfield Site Remediations" *Canadian Consulting Engineer* 51, no. 7: 25.

Cajete, G. 2000. *Native Science: Natural Laws of Interdependence*. Santa Fe, NM: Clear Light.

Callon, Michel. 1999. "The Role of Lay People in the Production and Dissemination of Scientific Knowledge." *Science, Technology and Society* 4, no. 1: 81–94.

Callon, Michel, Pierre Lascoumes, and Yannick Barthe. 2009. *Acting in an Uncertain World: An Essay on Technical Democracy*. Translated by Graham Burchell. Cambridge, MA: MIT Press.

Cameron, Emilie. 2012. "Securing Indigenous Politics: A Critique of the Vulnerability and Adaptation Approach to the Human Dimension of Climate Change in the Canadian Arctic." *Global Environmental Change* 22, no. 1: 103–14.

– 2015. *Far Off Metal River: Inuit Lands, Settler Stories, and the Making of the Contemporary Arctic*. Vancouver, BC: University of British Columbia Press.

Cameron, Emilie, Rebecca Mearns, and Janet Tamalik McGrath. 2015. "Translating Climate Change: Adaptation, Resilience and Climate Politics in Nunavut, Canada." *Annals of the Association of American Geographers* 105, no. 2: 274–83.

Campanella, David. 2015. *More Headaches Than It's Worth: Assessing Privatized and Semi-Privatized Waste Collection.* Ottawa, ON: Canadian Centre for Policy Alternatives.

Campbell, Brian L. 1985. "Uncertainty as Symbolic Action in Disputes among Experts." *Social Studies of Science* 15, no. 3: 429–53.

Canada–United States. 1955. *Establishment of a Distant Early Warning System: Agreement between Canada and the United States of America, Effected by an Exchange of Notes Signed at Washington, 5 May 1955.* Ottawa, ON: Canada Treaty Series, 1928–1964.

Canadian Institute for Environmental Law and Policy. 2008. *A Brief History of Waste Diversion in Ontario: A Background Paper on the Review of the Waste Diversion Act.* Toronto, ON: Canadian Institute for Environmental Law and Policy. http://www.cielap.org/pdf/WDA_BriefHistory.pdf.

Canadian Press. 2014. "Iqaluit's Long-Smouldering 'Dumpcano' Garbage Fire Finally Out." *Globe and Mail*, 16 September.

Capozza, K. L. 2002. "Ditched Drums and All." *Bulletin of the Atomic Scientists* 58: 14–6.

Carson, Rachel. 1962. *Silent Spring.* Boston, MA: Houghton Mifflin.

Castel, Robert. 1991. "From Dangerousness to Risk." In *the Foucault Effect: Studies in Governmentality*, edited by Graham Burchell, Colin Gordon, and Peter Miller, 281–98. Chicago, IL: University of Chicago Press

Cater, Tara Irene. 2013. "When Mining Comes (Back) to Town: Exploring Historical and Contemporary Mining Encounters in the Kivalliq Region, Nunavut." MA thesis, Memorial University of Newfoundland.

Catlin, Jesse R., and Yitong Wang. 2013. "Recycling Gone Bad: When the Option to Recycle Increases Resource Consumption." *Journal of Consumer Psychology* 23, 1: 122–7.

Cavell, Janice, and Jeff Noakes. 2010. *Acts of Occupation: Canada and Arctic Sovereignty, 1918–25.* Vancouver, BC: University of British Columbia Press.

Canadian Broadcasting Corporation (CBC) News. 2013a. "Canadians Produce More Garbage than Anyone Else," CBC News, 17 January.

– 2013b. "Feds Confirm Giant Mine Cleanup to Cost 1B." CBC News, 16 April.

– 2014a. "Nunavut Inuit Furious Over Seismic Testing Decision." CBC News North, 30 June.

– 2014b. "People with Breathing Issues Should Avoid Dump Smoke: Official." CBC News North, 23 May.

– 2014c. "Iqaluit #dumpcano Smoke Forces 2 Schools to Close." CBC News North, 6 June.
– 2014d. "Harper's Iqaluit Visit Prompts Protests over Food Costs, Dump Fire." CBC News North, 26 August.
– 2014e. "Iqaluit to Seek Military's Help in Tackling Dumpcano." CBC News North, 14 August.
– 2015. "Clyde River's Fight Against Seismic Testing in Federal Court." CBC News North, 20 April.
– 2016. "End of an Era for Nellie Cournoyea, the 'Iron Lady' of the North." CBC News North, 12 January.
Canadian Broadcasting Corporation (CBC) Radio *Current–* 2019a. "'Canada Is in the Wrong': Environmentalists Urge the Country to Clear Out Its Trash from the Philippines." CBC Radio, *The Current*, 25 April.
– 2019b. "'It's Pretty Staggering': Returned Online Purchases Often Sent to Landfill, Journalist's Research Reveals." CBC Radio, *The Current*, 12 December.
Canadian Broadcasting Corporation (CBC) Radio Q. 2014. "Tanya Tagaq Brings 'Animism' to Studio Q." CBC Radio Q. YouTube video, 16:50. Published 27 May. https://www.youtube.com/watch?v=ZuTIySphv2w.
Canadian Council of Academies (CCA). 2015. *Health Product Risk Communication: Is the Message Getting Through?* Ottawa, ON: The Expert Panel on the Effectiveness of Health Product Risk Communication, Council of Canadian Academies.
Celermajer, Danielle. 2009. *The Sins of the Nation and the Ritual of Apologies*. Cambridge, MA: Cambridge University Press.
Çelik, Başak, R. Kerry Rowe, and Kahraman Ünlü. 2009. "Effect of Vadose Zone on the Steady-State Leakage Rates from Landfill Barrier Systems." *Waste Management* 29, no. 1: 103–9.
Center for International Environmental Law. 2019. *Plastic and Climate: The Hidden Costs of a Plastic Planet*. Washington, DC: Center for International Environmental Law.
Center for Sustainability. 2010. "Problems with Current Recycling Methods." Grand Rapids, MI: Aquinas College. 21 June.
– n.d. "Sustainability Initiative at Aquinas College." Grand Rapids, MI: Aquinas College.
Chachamovich, Eduardo, Monica Tomlinson, Embrace Life Council, Nunavut Tunngavik, and the Government of Nunavut. 2013. *Learning from Lives that Have Been Lived: Nunavut Suicide Follow-Back Study 2005–2010*. Montreal, QC: Nunavut Suicide Follow-Back Study

Steering Committee and the McGill Group for Suicide Studies. https://assets.documentcloud.org/documents/708953/suicide-report-nunavut-english.pdf.

Chakrabarty, Dipesh. 2009. "The Climate of History: Four Theses." *Critical Inquiry* 35, no. 2: 197–222.

– 2012. "Postcolonial Studies and the Challenge of Climate Change." *New Literary History* 43, no. 1: 1–18.

– 2013. "History on an Expanded Canvas: The Anthropocene's Invitation." Presented at Anthropocene Project: An Opening. Berlin, Germany: 10–13 January. http://hkw.de/en/app/mediathek/video/22392.

Chandler, Alfred D. 1977. *The Visible Hand: The Managerial Revolution in American Business*. Cambridge, MA: Harvard University Press.

Chapman, Anne. 2004. "Technology as World Building." *Ethics, Place and Environment* 7, no. 1–2: 59–72.

Chertow, Marian. 2009. "The Ecology of Recycling." *UN Chronicle* 46, no. 3–4: 56–60.

Cheung, Shu Fai, Darius K-S Chan, and Zoe S-Y Wong. 1999. "Reexamining the Theory of Planned Behavior in Understanding Wastepaper Recycling." *Environment and Behavior* 31, no. 5: 587–612.

Chong, Wai K., and Christopher Hermreck. 2010. "Understanding Transportation Energy and Technical Metabolism of Construction Waste Recycling." *Resources, Conservation and Recycling* 54, no. 9: 579–90.

Chung, Chien-Ming. 2011. "China's E-Waste City" *Virginia Quarterly Review* 87, no. 2: 84–94.

City of Iqaluit. 2010. "General Plan: By-Law No. 703." Iqaluit, NU: City of Iqaluit. https://www.city.iqaluit.nu.ca/sites/default/files/iqaluit_general_plan_by-law_703_eng.pdf.

– 2014a. "About Iqaluit: Demographics." Iqaluit, NU: City of Iqaluit. https://www.city.iqaluit.nu.ca/visitors/explore-iqaluit/demographics.

– 2014b. "Iqaluit Sustainable Community Plan. Part 2: Action Plan." Iqaluit, NU: City of Iqaluit. https://www.city.iqaluit.nu.ca/content/iqaluit-sustainable-community-plan-archived.

City of Kingston. 2002. "Solid Waste Services Situation Analysis Review." Kingston, ON: City of Kingston.

– 2008. "City of Kingston and Kingston Census Metropolitan Area Statistical Profile." Kingston, ON: City of Kingston, Planning and Development.

– 2009. "Audited Financial Statements and Other Financial Information of the Corporation of the City of Kingston Year Ended December 31, 2009." Kingston, ON: City of Kingston.

- 2012a. "Audited Financial Statements and Other Financial Information of the Corporation of the City of Kingston Year Ended December 31, 2012." Kingston, ON: City of Kingston.
- 2012b. "Sustainability." Kingston, ON: City of Kingston.
- 2012c. "Environment, Infrastructure and Transportation Policies Committee Meeting #09-2012 Minutes for October 9." Kingston, ON: City of Kingston.
- 2013a. "City Hall Public Notice: City Wants to Recognize Remarkable Recyclers." News and Public Notices. Kingston, ON: City of Kingston.
- 2013b. "Environment, Infrastructure and Transportation Policies Committee Meeting #04-2013 Agenda for April 16." Kingston, ON: City of Kingston.
- 2013c. *Options to Maximize the Reduction, Reuse and Recycling of Kingston's Waste Stream.* Report no. EITP-13-002. Kingston, ON: City of Kingston.
- 2013d. "City Council Meeting no. 2013-03: Report to Council 13-051." Kingston, ON: City of Kingston.
- 2013e. *Waste Recycling Strategy 2010–2013 Volume III.* Report no. 13-186. Kingston, ON: City of Kingston, Solid Waste Division.
- 2013f. Environment, Infrastructure and Transportation Policies Committee. Meeting #04-2013, 16 April. Kingston, ON: City of Kingston.
- 2013g. "City Wants to Recognize Remarkable Recyclers." City Hall Public Notice, 18 March. Kingston, ON: City of Kingston. https://www.cityofkingston.ca/-/city-wants-to-recognize-remarkable-recyclers. Accessed 10 September 2013.
- 2014. "By-Law Number 2014-5: Solid Waste Management By-Law." Kingston, ON: City of Kingston. https://www.cityofkingston.ca/documents/10180/16904/Garbage+Bylaw.
- 2016. "Kingston's Strategic Plan 2015-2018." Kingston, ON: City of Kingston.
- 2019. *Waste Diversion Rate and Integrated Waste Management Plan.* Environment, Infrastructure and Transportation Policies Committee. Report #EITP-003. March 6. Kingston, ON: City of Kingston.

City of Whitehorse. n.d. "E-Waste." Whitehorse, YT: City of Whitehorse.

Clark, Nigel. 2005. "Ex-orbitant Globality." *Theory, Culture and Society* 22, no. 5: 165–85.
- 2008. "Aboriginal Cosmopolitanism." *International Journal of Urban and Regional Research* 32, no. 3: 737–44.
- 2010. "Ex-orbitant Generosity: Gifts of Love in a Cold Cosmos" *Parallax* 16, no. 1: 80–95.

– 2011. *Inhuman Nature: Sociable Life on a Dynamic Planet*. London, UK: Sage Publications.
Clark, Timothy. 2012. "Scale: Derangements of Scale." In *Telemorphosis: Theory in the Era of Climate Change*, vol. 1, edited by Tom Cohen, 148–66. Ann Arbor, MI: Open Humanities Press.
Code, Lorraine. 1991. *What Can She Know? Feminist Theory and the Construction of Knowledge*. Ithaca, NY: Cornell University Press.
– 1999. "Flourishing." *Ethics and the Environment* 4, no. 1: 63–72.
Collins, H. M., and Robert Evans. 2002. "The Third Wave of Science Studies: Studies of Expertise and Experience." *Social Studies of Science* 32, no. 2: 235–96.
Collins, Scott L., Fiorenza Micheli, and Laura Hartt. 2000. "A Method to Determine Rates and Patterns of Variability in Ecological Communities." *Oikos* 91, no. 2: 285–93.
Cone, Marla. 2005. "Rocket-Fuel Chemical Found in Breast Milk." *Los Angeles Times*. 23 February.
Conference Board of Canada. 2013. "Municipal Waste Generation." Ottawa, ON: Conference Board of Canada. http://www.conference-board.ca/hcp/details/environment/municipal-waste-generation.aspx.
Coninck, Pierre D., Michel Séguin, Esteban Chornet, Lucie Laramée, Astérie Twizeyemariya, Nicolas Abatzoglou, and Louis Racine. 1999. "Citizen Involvement in Waste Management: An Application of the STOPER Model via an Informed Consensus Approach." *Environmental Management* 23, no. 1: 87–94.
Contenta, Sandro. 2012. "DEW Line: Canada Is Cleaning Up Pollution Caused by Cold War Radar Stations in the Arctic." *The Star*, 4 August.
Cook, Aonghais, Steven Rushton, John Allan, and Andrew Baxter. 2008. "An Evaluation of Techniques to Control Problem Bird Species on Landfill Sites." *Environmental Management* 41, no. 6: 834–43.
Corporation of the Town of Iqaluit. "By-Law No. 537, Canadian Inuit Dog and Dog Team By-Law." Iqaluit, NU: Town of Iqaluit, 2001.
Corse, Ashley. 2012. "Nature as Infrastructure: Making and Managing the Panama Canal Watershed." *Social Studies of Science* 42, no. 4: 539–63.
Corvellec, Hervé. 2014. "Recycling Food Waste into Biogass, or How Management Transforms Overflows into Flow." In *Coping with Excess*, edited by Barbara Czarniawska and Orvar Löfren, 154–72. Cheltenham, UK: Edward Elgar Publishing.
– 2018. "Circular Economy Is Just Another Growth Model." Lund,

Sweden: Lund University. https://www.ism.lu.se/en/article/circular-economy-is-just-another-growth-model.

Corvellec, Hervé, and Johan Hultman. 2012. "From 'Less Landfilling' to 'Wasting Less': Societal Narratives, Socio-Materiality, and Organizations." *Journal of Organizational Change Management* 25, no. 2: 297–314.

Coulthard, Glen. 2014. *Red Skin, White Masks*. Minneapolis, MN: University of Minnesota Press.

Cronon, William. 1996. "The Trouble with Wilderness: Or, Getting Back to the Wrong Nature." *Environmental History* 1, no. 1: 7–28.

Crooks, Harold. 1993. *Giants of Garbage: The Rise of the Global Waste Industry and the Politics of Pollution Control*. Toronto, ON: Lorimer.

Croteau, Jean-Jacques. 2010. *Final Report of the Honorable Jean-Jacques Croteau Retired Judge of the Superior Regarding the Allegations Concerning the Slaughter of Inuit Sled Dogs in Nunavik (1950–1970)*. Quebec City, QC: Makivik Corporation and Government of Quebec.

Crutzen, Paul J., and Eugene F. Stoermer. 2000. "The Anthropocene." *Global Change Newsletter*, 41 (May): 17–8.

Damas, David. 2002. *Arctic Migrants, Arctic Villagers: The Transformation of Inuit Settlements in the Central Arctic*. Montreal and Kingston: McGill-Queen's University Press.

d'Anglure, Bernard and Levi-Strauss, Claude. 2018. *Inuit Stories of Being and Rebirth*. Winnipeg: University of Manitoba Press.

Danon-Schaffer, Monica. 2015. "Dumps, Landfills and Emerging Contaminants in the Canadian North." 2015 RPIC Federal Contaminated Sites Regional Workshop, Edmonton Alberta.

Darier, Éric. 1996a. "Environmental Governmentality: The Case of Canada's Green Plan." *Environmental Politics* 5, no. 4: 585–606.

– 1996b. "The Politics and Power Effects of Garbage Recycling in Halifax, Canada." *Local Environment* 1, no. 1: 63–86.

– 1999. *Discourses of the Environment*, Oxford, UK: Blackwell Publishers.

Dauvergne, Peter. 2010. *The Shadows of Consumption: Consequences for the Global Environment*. Cambridge, MA: MIT Press.

Davies, Anna R. 2008. *The Geographies of Garbage Governance: Interventions, Interactions and Outcomes*. Burlington: Ashgate Publishing Company.

Davies, Janette, Gordon R. Foxall, and John Pallister. 2002. "Beyond the Intention-Behaviour Mythology: An Integrated Model of Recycling." *Marketing Theory* 2, no. 1: 29–113.

Davis, Joseph. 2009. "Kingston: The Political Context of Brownfields in the City; Kingston as a Leader in Community Improvement Planning; Planning for Sustainability." In *The Story of Brownfields and Smart Growth in Kingston Ontario: From Contamination to Revitalization*, edited by Pamela Welbourn, Harry Cleghorn, Joseph Davis and Steven Rose, 85–92. Kingston: Classroom Complete Press.

Davis, Mike. 1996. "Cosmic Dancers on History's Stage? The Permanent Revolution in the Earth Sciences." *New Left Review*, no. 217: 48–84.

– 2007. *Planet of Slums*. New York: Verso.

Dean, Mitchell. 1999a. "Risk, Calculable and Incalculable." In *Risk and Sociocultural Theory: New Directions and Perspectives*, edited by Deborah Lupton, 131–59. Cambridge, MA: Cambridge University Press.

– 1999b. *Governmentality: Power and Rule in Modern Society*. London, UK: Sage Publications.

De Landa, Manuel. 1997. *A Thousand Years of Nonlinear History*. New York: Swerve.

Deleuze, Gilles, and Félix Guattari. 1980. *A Thousand Plateaus: Capitalism and Schizophrenia*. Translated by Brian Massumi. Minneapolis: University of Minnesota Press.

Deloitte. 2019. *Economic Study of the Canadian Plastic Industry, Market and Waste Task 5 – Summary Report to Environment and Climate Change Canada*. Deloitte LLC.

Department of Environmental Protection. 1998. *Gull Control at Landfills and Other Solid Waste Management Facilities*. Boston: Commonwealth of Massachusetts Executive Office of Environmental Affairs.

Department of Finance. 2014. *Towards a Representative Public Service: Statistics as of June 30th, 2014*. Iqaluit, NU: Government of Nunavut.

Department of Foreign Affairs and International Trade. 2010. *Statement on Canada's Arctic Foreign Policy: Exercising Sovereignty and Promoting Canada's Northern Strategy Abroad*. Written by Lawrence Cannon, Minister of Foreign Affairs. Ottawa, ON: Government of Canada.

Department of Health. 2014a. "Bulletin: Questions and Answers Iqaluit Dump Fire – Air Quality." Nunavut Department of Health, released 18 July 2014. https://www.gov.nu.ca/health/news/bulletin-questions-and-answers-iqaluit-dump-fire-air-quality.

– 2014b. "Bulletin: Questions and Answers Iqaluit Dump Fire – Air Quality." Nunavut Department of Health, released 31 July 2014. https://www.gov.nu.ca/health/news/bulletin-questions-and-answers-iqaluit-dump-fire-air-quality-update.

Department of Health and Social Services. 2007. *Nutrition in Nunavut: A Framework for Action*. Iqaluit, NU: Government of Nunavut. https://www.gov.nu.ca/sites/default/files/files/Nutrition%20Framework.pdf.

Department of Indian and Northern Affairs (DIAND). 2005. *Abandoned Military Site Protocol*. Ottawa, ON: Department of Indian and Northern Affairs.

Department of National Defence (DND). N.d. "Defence Environmental Strategy: A Plan for Ensuring Sustainable Military Operations." Ottawa, ON: Department of National Defence.

– 2014. "The Distant Early Warning (DEW) Line Remediation Project." Ottawa, ON: Department of National Defence. http://www.forces.gc.ca/en/news/article.page?doc=the-distant-early-dew-line-remediation-project/hgq87xvs

Derrida, Jacques. 1991. "'Eating Well,' or the Calculation of the Subject: An Interview with Jacques Derrida." In *Who Comes After the Subject?* edited by Eduardo Cadava, Peter Connor and Jean-Luc Nancy, 96–119. New York: Routledge.

– 1994. *Specters of Marx: The State of the Debt, the Work of Mourning, and the New International*. Translated by Peggy Kamuf. New York: Routledge.

– (1997) 2004. "The Animal That Therefore I Am (More to Follow)." In *Animal Philosophy: Essential Readings in Continental Thought*, edited by Matthew Calarco and Peter Atterton, 113–28. New York: Continuum.

– 2001a. "To Forgive: The Unforgivable and the Imprescriptible." In *Questioning God*, edited by John D. Caputo, Mark Dooley and Michael J. Scanlon, 21–51. Bloomington: Indiana University Press.

– 2001b. *On Cosmopolitanism and Forgiveness*. Translated by Mark Dooley and Michael Hughes. New York: Routledge.

DeLorenzo, A., K. Parizeau, and M. von Massow. 2019. "Regulating Ontario's Circular Economy through Food Waste Legislation." *Society and Business Review* 14, no. 2: 200–16.

DeSilvey, Caitlin. 2006. "Observed Decay: Telling Stories with Mutable Things." *Journal of Material Culture* 11, no. 3: 318–38.

DeSousa, Christopher A. 2006. "Urban Brownfields Redevelopment in Canada: The Role of Local Government." *Canadian Geographer* 50, no. 3: 392–407.

– 2003. "Turning Brownfields into Green Space in the City of Toronto." *Landscape and Urban Planning* 62, no. 4: 181–98.

De Vos, Rick. 2013. "Huskies and Hunters: Living and Dying in Arctic Greenland." In *Animal Death*, edited by Fiona Probyn-Rapsey and Jay Johnston, 277–92. Sydney, AU: Sydney University Press.

de Vries, Gerard. 2007. "What is Political in Sub-Politics? How Aristotle Might Help STS." *Social Studies of Science* 37, no. 5: 781–809.

Dewey, John. (1927) 1954. *The Public and its Problems: An Essay in Political Inquiry*. Athens: Swallow Press/Ohio University Press.

Dias, Sonia. 2016. "Waste Pickers and Cities." *Environment and Urbanization* 28, no. 2: 375–90.

Dias, Sonia and Lucia Fernandez. 2013. "Wastepickers: A Gendered Perspective." In *Powerful Synergies: Gender Equality, Economic Development and Environmental Sustainability*, edited by Blerta Cela, Irene Dankelman, and Jeffrey Stern. United Nations Development Programme, 153–5.

Dick, Terry A., Colin P. Gallagher, and Gregg T. Tomy. 2010. "Short-and Medium-Chain Chlorinated Paraffins in Fish, Water and Soils from the Iqaluit, Nunavut (Canada), Area." *World Review of Science, Technology and Sustainable Development* 7, no. 4: 387–401.

Dillon, Lindsey. 2014. "Race, Waste, and Space: Brownfield Redevelopment and Environmental Justice at the Hunters Point Shipyard." *Antipode* 46, no. 5: 1205–21.

Dillon Consulting. 2005. *Solid Waste Facility Operation and Maintenance Manual*. Iqaluit, NU: City of Iqaluit.

Di Menna, Jodi. 2006. "Giant Toxicle." *Canadian Geographic* 126, no. 2 (May/June): 30.

Diprose, Rosalyn. 2006. "Derrida and the Extraordinary Responsibility of Inheriting the Future-to-Come." *Social Semiotics* 16, no. 3: 435–47.

Dobbin, Terry. 2013. "Iqaluit Needs Help with Contaminated Site Cleanups." *Nunatsiaq News*, 14 January 2013. http://www.nunatsiaqonline.ca/stories/article/65674iqaluit_needs_help_with_contaminated_site_clean-ups/

Dodds, Klaus. 2012. "Graduated and Paternal Sovereignty: Stephen Harper, Operation NANOOK 10, and the Canadian Arctic." *Environment and Planning D: Society and Space* 30, no. 6: 989–1010.

Dodds, Lyn, and Bill Hopwood. 2006. "BAN Waste, Environmental Justice and Citizen Participation in Policy Setting." *Local Environment* 11, no. 3: 269–86.

Dolphijn, Rick, and Iris van der Tuin. 2012. *New Materialism: Interviews and Cartographies*. Ann Arbor, MI: Open Humanities Press.

Doron, Assa and Jeffrey, Robin. 2018. *Waste of a Nation: Garbage and Growth in India*. Cambridge, MA: Harvard University Press.

Douglas, Ian, and Nigel Lawson. 2000. "The Human Dimensions of Geomorphological Work in Britain." *Journal of Industrial Ecology* 4, no. 2: 9–33.

Douglas, Mary. 2007. *Purity and Danger: An Analysis of Concepts of Pollution and Taboo*. London and New York: Routledge.

Douglas, Mary, and Aaron Wildavsky. 1983. *Risk and Culture: An Essay on the Selection of Technological and Environmental Dangers*. Los Angeles: University of California Press.

Downs, Anthony. 1972. "Up and Down with Ecology: The Issue-Attention Cycle." *The Public Interest* 28 (Spring): 38–50.

Ducharme, Heather C. 2004. "Here We Fight the Coldest War: Environmental Science and Feminist Autobiography of the DEW Line." MA thesis, York University.

Durant, Darrin, and Genevieve F. Johnson, eds. 2009. *Nuclear Waste Management in Canada: Critical Issues, Critical Perspectives*. Vancouver, BC: University of British Columbia Press.

Duroy, Quentin. 2011. "The Path to a Sustainable Economy: Sustainable Consumption, Social Identity and Ecological Citizenship." *International Journal of Green Economics* 5, no. 1: 1–14.

Easterling, Keller. 2014. *Extrastatecraft: The Power of Infrastructure Space*. New York: Verso.

Eayrs, James. 1972. "In Defence of Canada: Peacemaking and Deterrence." Toronto, ON: University of Toronto Press.

EBA Engineering Consultants. 2009. *Comprehensive Solid Waste Study for Yukon Territory Waste Facilities*. Prepared for Government of Yukon, Department of Community Services.

Edmiston, Jake. 2010. "Don't Fear the Wrapper: Users Say Clear Bags No Biggie." *Kingston Whig-Standard*, 12 August.

Edwards, Tim. 2014. "Visions of Déline." *Up Here: Life in Canada's Far North* (July/August): 48–56.

Edwards, Katrina J., Keir Becker, and Frederick Colwell. 2012. "The Deep, Dark Energy Biosphere: Intraterrestrial Life on Earth." *Annual Review of Earth Planet Science* 40: 551–68.

Einsiedel, Edna F., Erling Jelsøe, and Thomas Breck. 2001. "Publics at the Technology Table: The Consensus Conference in Denmark, Canada, and Australia." *Public Understanding of Science* 10, no. 1: 83–98.

Eiselt, Horst A. 2006. "Locating Landfills: Optimization vs. Reality." *European Journal of Operational Research* 179, no. 3: 1,040–9.

Eisted, Rasmus, Anna W. Larsen, and Thomas H. Christensen. 2009.

"Collection, Transfer and Transport of Waste: Accounting of Greenhouse Gases and Global Warming Contribution." *Waste Management and Research* 27: 738–45.

Elagroudy, Sherien A., Mohamed H. Abdel-Razik, Mostafa A. Warith, and Fikry H. Ghobrial. 2008. "Waste Settlement in Bioreactor Landfill Models." *Waste Management* 28, no. 11: 2,366–74.

Elden, Stuart. 2013. "Secure the Volume: Vertical Geopolitics and the Depth of Power." *Political Geography* 34: 35–51.

Ellen TV. 2011. "Stop Seal Hunting in Canada Now." EllenTube. 6 April.

Elliott, Debbie. 2015. "5 Years after BP Oil Spill, Effects Linger and Recovery Is Slow." WBUR. 20 April. Accessed 4 August 2020.

Eno, Robert V. 2003. "Crystal Two: The Origin of Iqaluit." *Arctic* 56, no. 1: 63–75.

Environment Canada. 2005. "Legal Test for 'Deleterious Substance' Stands." *Ontario Region COMPRO Update* 11, no. 1. Accessed 14 September 2012. http://www.on.ec.gc.ca/epb/fpd/compro/2020-e.pdf.

– 2013. "Managing and Reducing Waste." Ottawa, ON: Government of Canada. http://www.ec.gc.ca/gdd-mw.

Environment and Climate Change Canada. 2017. *Solid Waste Management for Northern and Remote Communities: Planning and Technical Guidance Document*. Ottawa, ON: Government of Canada.

– 2018. *Canadian Environmental and Sustainability Indicators: Solid Waste Diversion and Disposal*. Ottawa, ON: Government of Canada.

Environmental Commissioner of Ontario. 2017. *Beyond the Blue Box: Ontario's Fresh Start on Waste Diversion and the Circular Economy*. Toronto, ON: Environmental Commissioner of Ontario.

Environmental Protection Agency. 2012. "Brownfields." Washington, DC: United States Environmental Protection Agency. http://www.epa.gov/brownfields.

Environmental Sciences Group (ESG). 1995. *Environmental Study of a Military Installation and Six Waste Disposal Sites at Iqaluit, NWT*, vol. 1: Site Analysis. Ottawa, ON: Indian and Northern Affairs Canada and Environment Canada.

Environmental Sciences Group (ESG) and UMA Engineering. 1995. *DEW Line Cleanup: Scientific and Engineering Summary Report*. Ontawa, ON: Minister of National Defence.

European Commission. 2008. "Directive 2008/98/EC of the European Parliament and of the Council of 19 November 2008 on Waste and Repealing Certain Directives" (OJ L 312), Brussels, Belgium: European Commission, 3–30.

– 2014. "Trade in Seal Products: Scope of the EU Seal Ban." Brussels, Belgium: European Commission. Accessed 11 January 2014. http://ec.europa.eu/environment/biodiversity/animal_welfare/seals/seal_hunting.htm.

Ewald, Francois. 1991. "Insurance and Risk." In *the Foucault Effect: Studies in Governmentality*, edited by Graham Burchell, Colin Gordon, and Peter Miller, 197–210. Chicago, IL: University of Chicago Press.

Exp Services. 2011. *Iqaluit Waste Management Project: Preliminary Brief on Diversion and Disposal Options*. Project No. OTT-00020728. Prepared for City of Iqaluit. https://iqaluitwasteproject.files.wordpress.com/2011/06/iqaluit-waste-project-options-brief_june-20.pdf.

– 2013. *City of Iqaluit Solid Waste Management Plan*. Prepared for City of Iqaluit. https://iqaluitwasteproject.files.wordpress.com/2013/12/draft-solid-waste-management-plan_english-2013.pdf.

Fahnestock, Jeanne, and Marie Secor. 1998. "The Stases in Scientific and Literary Argument." *Written Communication* 5, no. 4: 427–43.

Farish, Matthew. 2006. "Frontier Engineering: From the Globe to the Body in the Cold War Arctic." *Canadian Geographer* 50, no. 2: 177–94.

Farish, Matthew, and P. Whitney Lackenbauer. 2009. "High Modernism in the Arctic: Planning Frobisher Bay and Inuvik." *Journal of Historical Geography* 35, no. 3: 517–44.

Fausto-Sterling, Anne. 1997. "Feminism and Behavioral Evolution: A Taxonomy." In *Feminism and Evolutionary Biology*, edited by Patty Gowaty, 42–60. New York: Chapman and Hall.

Federal Emergency Management Agency (FEMA). 2002. *Landfill Fires: Their Magnitude, Characteristics, and Mitigation*. Arlington, VA: United States Fire Administration, National Fire Data Centre.

Ferguson, Elliott. 2011. "Ministry Fines Norterra." *Kingston Whig-Standard*. 10 May.

Field, Hyman, and Patricia Powell. 2001. "Public Understanding of Science versus Public Understanding of Research." *Public Understanding of Science* 10: 421–6.

Fleck, Ludwik. 1979. *Genesis and Development of a Scientific Fact*. Chicago, IL: University of Chicago Press.

Foote, Stephanie, and Elizabeth Mazzolini. 2012. *Histories of the Dustheap: Waste, Material Cultures, Social Justice*. Cambridge, MA: MIT Press.

Foucault, Michel. 1980. "Two Lectures." In *Power/Knowledge: Selected Interviews and Other Writings 1972–1977*, edited by Colin Gordon, 78–108. New York: Pantheon Books.

– 1984a. "Nietzsche, Genealogy, History." In *The Foucault Reader*, edited by Paul Rabinow, 77–100. New York: Pantheon Books.
– 1984b. "The Politics of Health in the Eighteenth Century." In *The Foucault Reader*, edited by Paul Rabinow, 273–89. New York: Pantheon Books.
– 1988. "Technologies of the Self." In *Technologies of the Self: A Seminar with Michel Foucault*, edited by Luther H. Martin, Huck Gutman, and Patrick H. Hutton, 16–49. London, UK: Tavistock Publications.
– 1991. "Governmentality." In *The Foucault Effect: Studies in Governmentality*, edited by Graham Burchell, Colin Gordon, and Peter Miller, 87–104. Chicago, IL: University of Chicago Press.
Fox Keller, Evelyn. 1983. *A Feeling for the Organism: The Life and Work of Barbara McClintock*. San Francisco, CA: W. H. Freeman.
Franklin, Adrian. 2008. "A Choreography of Fire: A Posthumanist Account of Australians and Eucalypts." In *The Mangle in Practice: Science, Society, and Becoming*, edited by Andrew Pickering and Keith Guzik, 17–45. Durham, NC: Duke University Press.
Franz Environmental. 2008. *Phase I/II Environmental Site Assessment: Vehicle Dump and Community Landfill, Iqaluit, Nunavut*. Vancouver, BC: Public Works and Government Services Canada.
Freeman, John. 2015. "Breath of the Inuit: Tanya Tagaq Interviewed." *Quietus*, 28 January.
Freud, Sigmund. 2010. *Civilization and its Discontents*. Eastford, CT: Martino Publishing.
Fried, Jacob. 1963. "Settlement Types and Community Organization in Northern Canada." *Arctic* 16, no. 2: 93–100.
Friesen, Joe. 2009. "How the Little People Stopped the Tiny Township Dump." *Globe and Mail*, 28 August.
Frodeman, Robert. 2003. *Geo-Logic: Breaking Ground between Philosophy and the Earth Sciences*. New York: State University of New York Press.
Fuller, Steve. 1987. "Social Epistemology: A Statement of Purpose." *Social Epistemology* 1, no. 1: 1–4.
Furedy, Christine. 1993. *Solid Waste in the Waste Economy: Socio-Cultural Aspects*. Unpublished paper, York University. http://www.yorku.ca/furedy/papers/hanpap.htm. Accessed 5 August 2020.
Gabrys, Jennifer. 2007. "Media in the Dump." In *Alphabet City: Trash*, no. 11, edited by John Knechtel, 156–65. Cambridge, MA: MIT Press.
– 2011. *Digital Rubbish: A Natural History of Electronics*. Ann Arbor, MI: University of Michigan Press.

– 2013. "Plastic and the Work of the Biodegradable." In *Accumulation: The Material Politics of Plastic*, edited by Jennifer Gabrys, Gay Hawkins, and Mike Michael, 208–27. Oxon, UK: Routledge.

Gagnon, Mélanie, and Iqaluit Elders. 2002. *Inuit Recollections on the Military Presence in Iqaluit*. Prepared for Nunavut Arctic College.

Galison, Peter, and Jamie Kruse. 2014. "Waste-Wilderness: A Conversation between Peter Galison and Smudge Studio." *Discard Studies*, 26 March. http://discardstudies.com/2014/03/26/waste-wilderness-a-conversation-between-peter-galison-and-smudge-studio.

Gallagher, Louise, Susana Ferreira, and Frank Convery. 2008. "Host Community Attitudes towards Solid Waste Landfill Infrastructure: Comprehension before Compensation." *Journal of Environmental Planning and Management* 51, no. 2: 233–57.

Gartner Lee. 2007. *City of Yellowknife Solid Waste Composition Study and Waste Reduction Recommendations*. Prepared for City of Yellowknife.

George, Jane. 2007. "Pang Residents All Choked Up over Trash Fire: 'It Smelled Like a Dump, like Chemicals.'" *Nunatsiaq News*, 27 July.

Gerrard, Michael B. 1995. "Dodging the NIMBY Bullet: A Solution to Waste Facility Siting." *Public Utilities Fortnightly* 133, no. 15 (August): 18–9.

Gharfalkar, Mangesh, Richard Court, Callum Campbell, Zulfiqur Ali, and Graham Hillier. 2015. "Analysis of Waste Hierarchy in the European Waste Directive 2008/98/EC." *Waste Management* 39: 305–13.

Ghoreishi, Omid. 2013. "Canada Highest Per Capita Waste Producer Compared to Other Developed Nations." *Epoch Times*, 30 January–5 February: 8.

Ghosn, Rania, and El Hadi Jazairy. 2015. *Geographies of Trash*. Barcelona and New York: Actar Publishers.

Giddens, Anthony. 1990. *The Consequences of Modernity*. Cambridge, MA: Polity Press.

Giesy, John P., and Kurunthachalam Kannan. 1998. "Dioxin-like and Non-Dioxin-like Toxic Effects of Polychlorinated Biphenyls (PCBs): Implications for Risk Assessment." *Critical Reviews in Toxicology* 28, no. 6: 511–69.

Gilg, Andrew, Stewart Barr, and Nicholas Ford. 2005. "Green Consumption or Sustainable Lifestyles? Identifying the Sustainable Consumer." *Futures* 37, no. 6: 481–504.

Gille, Zsuzsa. 2007. *From the Cult of Waste to the Trash Heap of History: The Politics of Waste in Socialist and Postsocialist Hungary*. Bloomington: Indiana University Press.

– 2010. "Actor Networks, Modes of Production, and Waste Regimes: Reassembling the Macro-social." *Environment and Planning A* 42, no. 5: 1,049–64.

*Globe and Mail*. 1998. "City Guilty of Polluting River." *Globe and Mail*, Environmental Bureau of Investigation, 12 December.

Goodman, Lee-Anne. 2015. "50 Containers of Canadian Garbage Rots in Manila for Two Years." *The Star*. 20 March.

Gordillo, Gastón R. 2014. *Rubble: The Afterlife of Destruction*. Durham, NC: Duke University Press.

Gordon, Colin. 1991. "Governmental Rationality: An Introduction." In *The Foucault Effect: Studies in Governmentality*, edited by Graham Burchell, Colin Gordon, and Peter Miller, 1–52. Chicago, IL: University of Chicago Press.

Gordon, James. 2014. "Dumpcano and Canada's Northern Hypocrisy." *Ottawa Citizen*, 7 August.

Goven, Joanna. 2003. "Deploying the Consensus Conference in New Zealand: Democracy and De-Problematization." *Public Understanding of Science* 12, no. 4: 423–40.

– 2006. "Dialogue, Governance, and Biotechnology: Acknowledging the Context of the Conversation." *Integrated Assessment Journal* 6, no. 2: 99–116.

Government of Canada. 2010. "High Investment Potential in Canadian Northern Oil and Gas." Ottawa, ON: Indigenous and Northern Affairs Canada. https://www.aadnc-aandc.gc.ca/eng/1100100037174/1100100 037175.

– 2014. "Canada's Northern Strategy." Ottawa, ON: Government of Canada. http://www.northernstrategy.gc.ca/index-eng.asp.

Government of Canada. 2016. "Municipal Solid Waste and Greenhouse Gases." https://www.canada.ca/en/environment-climate-change/services/ managing-reducing-waste/municipal-solid/greenhouse-gases.html.

Government of Northwest Territories. 1972. *Report: Environment Frobisher Bay*. Iqaluit, NWT: City of Iqaluit.

Government of Nunavut. 2014a. *Air Pollution Concentrations in Iqaluit, Nunavut (June 14–September 22, 2014)*. Ottawa, ON: Water and Air Quality Bureau, Health Canada.

– 2014b. "Public Health Advisory: Iqaluit Dump Fire Air Quality Testing Update." Environmental Registry. Toronto, ON: Government of Ontario.

Government of Ontario. 2002. *Waste Diversion Act 2002*. Toronto, ON: Government of Ontario.

Government of Yukon. 2012. "Yukon Moves Toward Modern Solid Waste Management System." Whitehorse, YT: Government of Yukon. https://open.yukon.ca/data/sites/default/files/20120309YTMovesTowardModernSolidWasteManagementSystem.pdf.

Graham, Stephen, and Nigel Thrift. 2007. "Out of Order: Understanding Repair and Maintenance." *Theory, Culture and Society* 24, no. 3: 1–25.

Grajeda, Tony. 2005. "Disasterologies." *Social Epistemology* 19, no. 4: 315–19.

Grasswick, Heidi E., and Mark Owen Webb. 2002. "Feminist Epistemology as Social Epistemology." *Social Epistemology* 16, no. 3: 185–96.

Gregory, Derek. 2006. *The Colonial Present*. Malden, MA: Blackwell.

Gregson, Nicky, and Mike Crang. 2010. "Guest Editorial. Materiality and Waste: Inorganic Vitality in a Networked World." *Environment and Planning A* 42, no. 5: 1026–32.

Gregson, Nicky, Helen Watkins, and Melania Calestani. 2010. "Inextinguishable Fibres: Demolition and the Vital Materialisms of Asbestos." *Environment and Planning A* 42, no. 5: 1065–83.

Griffiths, Franklyn. 2011. "Towards a Canadian Arctic Strategy." In *Canada and the Changing Arctic: Sovereignty, Security, and Stewardship*, edited by Franklyn Griffiths, Rob Huebert, and P. Whitney Lackenbauer, 181–226. Waterloo, ON: Wilfrid Laurier University Press.

Griffiths, Franklyn, Rob Huebert, and P. Whitney Lackenbauer, eds. 2011. *Canada and the Changing Arctic: Sovereignty, Security, and Stewardship*. Waterloo, ON: Wilfrid Laurier University Press.

Grover, Velma I. 2010. "Brownfields." *Sage Knowledge Encyclopedia*. Thousand Oaks, CA: Sage Publishing: 1–3.

Hacking, Ian. 1999. *The Social Construction of What?* Cambridge, MA: Harvard University Press.

Hale, Benjamin, and Lisa Dilling 2011. "Geoengineering, Ocean Fertilization, and the Problem of Permissible Pollution." *Science, Technology, and Human Values* 36, no. 2: 190–212.

Hankinson-Nelson, Lynn, and Jack Nelson. 1996. *Feminism, Science, and the Philosophy of Science*. Dordrecht, Netherlands: Kluwer Academic Publishers.

Hannigan, John. 2006. *Environmental Sociology*, 2nd edition. London and New York: Routledge.

Haraway, Donna. 1988. "Situated Knowledges: The Science Question in Feminism and the Privilege of Partial Perspective." *Feminist Studies* 14, no. 3: 575–99.

– 1989. *Primate Visions: Gender, Race, and Nature in the World of Modern Science*. New York and London: Routledge.
– 1991. *Simians, Cyborgs, and Women: The Reinvention of Nature*. New York and London: Routledge.
– 1997. *Modest_Witness@Second_Millennium. FemaleManÓ_Meets_ OncoMouseÓ*. New York and London: Routledge.
– 2008. *When Species Meet*. Minneapolis, MN: University of Minnesota Press.
Hargreaves, Tom. 2012. "Questioning the Virtues of Pro-Environmental Behaviour Research: Towards a Phronetic Approach." *Geoforum* 43, no. 2: 315–24.
Harman, Graham. 2007. "On Vicarious Causation." In *Collapse Volume II: Speculative Realism*, edited by Robin Mackay and Damian Veal, 187–221. Falmouth, UK: Urbanomic.
– 2008. "The Metaphysics of Objects: Latour and his Aftermath." Draft paper, American University in Cairo. https://pervegalit.files.wordpress.com/2008/06/harmangraham-latour.pdf.
– 2010a. "Asymmetrical Causation: Influence without Recompense." *Parallax* 16: 96–109.
– 2010b. "I Am Also of the Opinion That Materialism Must Be Destroyed." *Environment and Planning D: Society and Space* 28, no. 5: 772–90.
– 2011a. "Autonomous Objects." *New Formations*, 71: 125–30.
– 2011b. *The Quadruple Object*. Winchester, UK: Zero Books.
– 2013. "The Current State of Speculative Realism." *Speculations: A Journal of Speculative Realism* 4: 22–8.
Harper, Stephen. 2013. "Speech from the Throne to Open the Second Session Forty First Parliament of Canada." Ottawa, ON: Parliament of Canada, Parlinfo. https://lop.parl.ca/sites/ParlInfo/default/en_CA/Parliament/procedure/throneSpeech/speech412.
Harris, Francis. 2005. "Canada Flexes its Muscles in Dispute over Arctic Wastes." *Telegraph*, 22 August.
Harris, Lynden T. n.d. "The DEW Line Chronicles: A History." The DEW-Line. Accessed 4 January 2015. http://lswilson.dewlineadventures.com/dewhist-a.
Harris, Sophia. 2015. "Canadians Piling Up More Garbage than Ever Before as Disposables Rule." CBC News.
Harrison, Phyllis, ed. 1964. *Q-Book: Qaujivaallirutissat*. Prepared for Department of Northern Affairs and Northern Development, Ottawa,

ON. http://publications.gc.ca/collections/collection_2018/aanc-inac/R72-3463.pdf.
Hartnett, Carolyn G. 2004. "How Does Science Express Uncertainty?" *LACUS Forum* 30: 355–65.
Hawkins, Gay. 2006. *The Ethics of Waste: How We Relate to Rubbish*. London, UK: Rowman and Littlefield Publishers.
– 2011. "Packaging Water: Plastic Bottles as Market and Public Devices." *Economy and Society* 40, no. 4: 534–52.
Hawkins, Gay, Emily Potter, and Kane Race. 2015. *Plastic Water: The Social and Material Life of Bottled Water*. Cambridge, MA: MIT Press.
Hazen, Terry C., Eric A. Dubinsky, Todd Z. DeSantis, Gary L. Andersen, Yvette M. Piceno, Navjeet Singh, Janet K. Jansson, et al. 2010. "Deep-Sea Oil Plume Enriches Indigenous Oil-Degrading Bacteria." *Science* 330, no. 6,001: 204–8.
HDR Corporation. 2011. *Investigation of Residual Waste Processing Systems*. Report prepared for City of Kingston, Project no. 172826. https://www.cityofkingston.ca/documents/10180/984482/EIT_A0912-12023.pdf/ea5bdf20-538d-4735-8145-2a937d53680f.
Health Canada. 2005. "Addressing Psychosocial Factors through Capacity Building: A Guide for Managers of Contaminated Sites." Ottawa, ON: Government of Canada, Catalogue no. H46-2/05-430E. Ottawa, ON: Queen's Printer. https://www.canada.ca/content/dam/hc-sc/migration/hc-sc/ewh-semt/alt_formats/hecs-sesc/pdf/pubs/contamsite/guide/guide-eng.pdf.
– 2006. "Dioxins and Furans." *Health Canada*. http://www.hc-sc.gc.ca/hl-vs/iyh-vsv/environ/dioxin-eng.php.
– 2010. "A Guide to Involving Aboriginal Peoples in Contaminated Site Management." Catalogue no. H128-1/10-628E-PDF. Ottawa, ON: Queen's Printer. http://publications.gc.ca/pub?id=9.694028&sl=0.
– 2014. "Air Pollution Concentrations in Iqaluit, Nunavut (June 14 – September 22, 2014)." Prepared for the Government of Nunavut. Water and Air Quality Bureau, Health Canada. http://www.gov.nu.ca/sites/default/files/files/Iqaluit%20Air%20Monitoring%20Data%20Report%20June%202014%20-%20Sept%2022,%202014.pdf.
Healy, Stephen A. 2010. "Facilitating Public Participation in Toxic Waste Management Through Engaging 'The Object of Politics.'" *East Asian Science, Technology and Society: An International Journal* 4, no. 4: 585–99.
Heaps, Leo. 1978. *Operation Morning Light: Terror in Our Skies: The True Story of Cosmos 954*. New York: Paddington Press.

Heidegger, Martin. 1991. *Nietzsche: Volumes Three and Four*, edited by David Farrell Krell. San Francisco, CA: HarperOne.

Heidt, Dan, and P. Whitney Lackenbauer. 2012. "Sovereignty for Hire: Civilian Airlift Contractors and the Distant Early Warning (DEW) Line, 1954–1961." *Canadian Aerospace Power Studies* 4: 95–112.

Heimer, Carol A. 1985. *Reactive Risk and Rational Action: Managing Moral Hazard in Insurance Contracts*. Berkeley, CA: University of California Press.

Heinke, Gary W., and Jeffrey Wong. 1990. *Solid Waste Composition Study for Iqaluit, Pangnirtung, and Broughton Island of the Northwest Territories*. Prepared for Department of Municipal and Community Affairs Government of the Northwest Territories.

Henri, Dominique. 2012. "Managing Nature, Producing Cultures: Inuit Participation, Science and Policy in Wildlife Governance in the Nunavut Territory, Canada." PhD thesis, Oxford University.

Hetherington, Kevin. 2004. "Secondhandedness: Consumption, Disposal, and Absent Presence." *Environment and Planning D: Society and Space* 22, no. 1: 157–73.

Hird, Myra J. 2004. *Sex, Gender and Science*. Houndmills, Basingstoke, UK: Palgrave Press.

– 2009. *The Origins of Sociable Life: Evolution After Science Studies*. Houndmills, Basingstoke, UK: Palgrave Press.

– 2010a. "Indifferent Globality." *Theory, Culture and Society* 27, no. 2–3: 54–72.

– 2010b. "Meeting with the Microcosmos." *Environment and Planning D: Society and Space* 28, no. 1: 36–9.

– 2010c. "Coevolution, Symbiosis, and Sociology." *Ecological Economics* 69, no. 4: 737–42.

– 2010d. "The Life of the Gift." *Parallax* 16, no. 1: 1–6.

– 2012a. "Knowing Waste: Towards an Inhuman Epistemology." *Social Epistemology* 26, no. 3–4: 453–69.

– 2012b. "Wasteflow, Care, and an Ethic of Vulnerability." Unpublished paper.

– 2012c "Animal, All Too Animal: Toward an Ethic of Vulnerability." In *Animal Others and the Human Imagination*, edited by Aaron Gross and Anne Vallely, 331–48. New York: Columbia University Press.

– 2013. "Is Waste Indeterminacy Useful?: A Response to Zsuzsa Gille." *Social Epistemology Review and Reply Collective* 2, no. 6: 28–33.

– 2017. "Proliferation, Extinction, and the Anthropocene Aesthetic." In *Posthumous Life: Theorizing Beyond the Posthuman*, edited by Jami

Weinstein and Claire Colebrook, 251–70. New York: Columbia University Press.

Hird, Myra J., Scott Lougheed, R. Kerry Rowe, and Cassandra Kuyvenhoven. 2014. "Making Waste Management Public (or Falling Back to Sleep)." *Social Studies of Science* 44, no. 3: 441–65.

Hobson, Kersty. 2006. "Bins, Bulbs, and Shower Timers: On the 'Techno-Ethics' of Sustainable Living." *Ethics, Place and Environment* 9, no. 3: 317–36.

– 2013. "On the Making of the Environmental Citizen." *Environmental Politics* 22, no. 1: 56–72.

Hobson, Kersty, and Ann Hill. 2010. "Cultivating Citizen-Subjects through Collective Praxis: Organized Gardening Projects in Australia and the Philippines." In *Ethical Consumption: A Critical Introduction*, edited by Tania Lewis and Emily Potter, 216–30. London, UK: Routledge.

Honigmann, John J., and Irma Honigmann. 1965. "How Baffin Island Eskimo Have Learned to Use Alcohol." *Social Forces* 44, no. 1: 73–83.

Hooks, Gregory, and Chad Smith. 2004. "The Treadmill of Destruction." *American Sociological Review* 69, no. 4: 558–75.

Hopper, Tristin. 2014. "Mountain of Vancouver Garbage That Ended Up in Manila Has Philippines Demanding Canada Repatriate Its 'Junk,'" *National Post*. 15 October.

– 2015. "Years after 2,500 Tonnes of Canadian Trash Landed in Manila, Philippines Demanding We Take It Back," *National Post*. 12 October.

Hossay, Patrick 2006. *Unsustainable: A Primer for Global Environmental and Social Justice*. London, UK: Zed Books.

Hough Woodland Naylor Dance Leinster, Duke Engineering Services (Canada) Angus Environmental, Michael Michalski Associates, and DS-Lea Associates. 1998. *The Arsenal Lands: Park and Site Remediation Master Plan*. Prepared for the Toronto and Region Conservation Authority.

Hubbard, Ruth. 1989. "Science, Facts, and Feminism." In *Feminism and Science*, edited by Nancy Tuana, 119–31. Bloomington: Indiana University Press.

Huebert, Rob. 2011. "Canadian Arctic Sovereignty and Security in a Transforming Circumpolar World." In *Canada and the Changing Arctic: Sovereignty, Security, and Stewardship*, edited by Franklyn Griffiths, Rob Huebert, and P. Whitney Lackenbauer, 13–68. Waterloo, ON: Wilfrid Laurier University Press.

Hultman, Johan, and Hervé Corvellec. 2012. "The European Waste

Hierarchy: From the Sociomateriality of Waste to a Politics of Consumption." *Environment and Planning A* 44: 2,413–27.

Humes, Edward. 2012. *Garbology: Our Dirty Love Affair with Trash.* New York: Avery Press.

Hunt, Jane, and Brian Wynne. 2000. *Forums for Dialogue: Developing Legitimate Authority through Communication and Consultation.* A contract report for Nirex. Lancaster, UK: Centre for the Study of Environmental Change, Lancaster University.

Inuit Circumpolar Council (ICC). 2009a. *A Circumpolar Declaration on Sovereignty in the Arctic.* Anchorage, AK: Inuit Circumpolar Council.

– 2009b. *A Circumpolar Inuit Declaration on Resource Development Principles in Inuit Nunaat.* Anchorage, AK: Inuit Circumpolar Council.

Iqalummiut for Action (IFA). 2014. "An Open Letter Regarding Operation NANOOK and the Iqaluit Dump Fire." Letter submitted to Honourable Robert Nicholson and General Thomas J. Lawson. Iqaluit, NU: Iqalummiut for Action, 8 August.

Indian and Northern Affairs Canada. 2008. *Giant Mine Remediation Project: Moving Forward Together.* Ottawa, ON: Indian and Northern Affairs Canada.

Ingold, Tim. 2007. "Materials Against Materiality." *Archaeological Dialogues* 14, no. 1: 1–16.

Insurance Institute for Highway Safety. 2016. "Highway Loss Data: General Statistics." Arlington, VA: Insurance Institute for Highway Safety.

Irwin, Alan. 2008. "Risk, Science and Public Communication: Third-Order Thinking about Scientific Culture." In *Handbook of Public Communication of Science and Technology,* edited by Massimiano Bucchi and Brian Trench, 199–212. New York: Routledge.

Irwin, Alan, and Brian Wynne. 1996. "Introduction." In *Misunderstanding Science? The Public Reconstruction of Science and Technology,* edited by Alan Irwin and Brian Wynne, 1–18. Cambridge, UK: Cambridge University Press.

Iryo, Takamasa, and R. Kerry Rowe. 2003. "On the Hydraulic Behaviour of Unsaturated Nonwoven Geotextiles." *Geotextiles and Geomembranes* 21, no. 6: 381–404.

– 2005. "Hydraulic Behaviour of Soil: Geocomposite Layers in Slopes." *Geosynthetics International* 12, no. 3: 145–55.

Islam, Mohammad Zahidul, and R. Kerry Rowe. 2009. "Permeation of BTEX through Unaged and Aged HDPE Geomembranes." *ASCE Journal of Geotechnical and Geoenvironmental Engineering* 135, no. 8: 1130–40.

Jacques Whitford. 2008a. *Phase A Report: Integrated Waste Management Study, the City of Kingston*. Report for the City of Kingston, Project No. 1018165. https://www.cityofkingston.ca/documents/10180/27835/Study_IntegratedWasteManagementStudy.pdf.

– 2008b. *Phase B Report: Integrated Waste Management Study, The City of Kingston. Report for the City of Kingston*. Report for the City of Kingston, Project No. 1018165. Accessed 10 March 2014. http://www.cityofkingston.ca/pdf/waste/IntegratedWasteManagementStudy_PhaseB.pdf.

Jasanoff, Sheila. 2003. "Breaking the Waves in Science Studies: Comment on H. M. Collins and Robert Evans, 'The Third Wave of Science Studies.'" *Social Studies of Science* 33, no. 3: 389–400.

Jasanoff, Sheila, ed. 2004. *States of Knowledge: The Co-Production of Science and Social Order*. London, UK: Routledge.

Jay, Dru Oja. 2013. "What If Natives Stop Subsidizing Canada?" Media Co-op, 7 January. http://www.mediacoop.ca/blog/dru/15493.

Jenness, Diamond. 1959. *The People of the Twilight*. Chicago, IL: University of Chicago Press.

Johnson, Genevieve. 2008. *Deliberative Democracy for the Future: The Case of Nuclear Waste Management in Canada*. Toronto, ON: University of Toronto Press.

Johansson, Nils. 2013. *Why Don't We Mine the Landfills?* Linköping, Sweden: Linköping Electronic Press.

Johansson, Nils, and Jonathan Metzger. 2014. "Experimentalizing the Organization of Objects: Some Why's and How's Discussed in Relation to Mineral Resource Becomings." Unpublished paper.

John, Robert. 2012. "City of Kingston to Recognize 'Remarkable Recyclers.'" *Kingston Herald*, 29 February.

Johnson, Genevieve F. 2008. *Deliberative Democracy for the Future: The Case of Nuclear Waste Management in Canada*. Toronto, ON: University of Toronto Press.

Johnson, Ken. 2001. "Land Use Planning and Waste Management in Iqaluit, Nunavut." *2001: A Spatial Odyssey/Odyssé de l'espace*: 124–7.

– 2005. "Social and Infrastructure Impacts of Landfill Sites in Cold Region Communities." Paper presented at Waste: The Social Context, Edmonton, AB, 11–14 May.

Johnson, Robert. 2009. "Wind, Wings and Waste." *Waste 360*, 1 June. http://waste360.com/Landfill_Management/managing-birds-blown-landfill-litter-200906.

Joseph, Sean. 2012. Email to Mike Molinski of Transport Canada. Iqaluit, NU: Nunavut Impact Review Board. Accessed 16 May 2015.

Judd, Alexander. 2005. *In Defense of Garbage*. Westport and London, UK: Praeger Publishers.

Kafarowski, Joanna. 2004. "Gender, Culture, and Contaminants in the North." *Signs* 34, no. 3: 494–9.

Kahhat, Ramzy, and Eric Williams. 2012. "Materials Flow Analysis of E-Waste: Domestic Flows and Exports of Used Computers from the United States." *Resources, Conservation and Recycling* 67: 67–74.

Kahrilas, Genevieve A., Jens Blotevogel, Philip S. Stewart, and Thomas Borch. 2015. "Biocides in Hydraulic Fracturing Fluids: A Critical Review of Their Usage, Mobility, Degradation, and Toxicity." *Environmental Science and Technology* 49, no. 1: 16–32.

Kalinovich, Indra, Allison Rutter, John S. Poland, Graham Cairns, and R. Kerry Rowe. 2008. "Remediation of PCB Contaminated Soils in the Canadian Arctic: Excavation and Surface PRB Technology." *Science of the Total Environment* 407, no. 1: 53–66.

Kant, Immanuel. 2003. *Observations on the Feeling of the Beautiful and Sublime*. Translated by John T. Goldthwait. Los Angeles, CA: University of California Press.

Kapur, Amit, and Thomas E. Graedel. 2006. "Copper Mines Above and Below the Ground." *Environmental Science and Technology* 40, no. 10: 3135–41.

Karidis, Arlene. 2016. "Rising Landfill Temperatures Across Country Spark Speculation Among Officials." *Waste Dive*. 28 March. http://www.wastedive.com/news/rising-landfill-temperatures-across-country-spark-speculation-among-officia/416301.

Kärrholm, Mattias. 2013. "Building Type Production and Everyday Life: Rethinking Building Types through Actor-Network Theory and Object-Oriented Philosophy." *Environment and Planning D: Society and Space* 31, no. 6: 1109–24.

Katz, Cheryl. 2012. "Melting Glaciers Liberate Ancient Microbes." *Scientific American*, 18 April.

Keeling, Arn, and John Sandlos. 2009. "Environmental Justice Goes Underground? Historical Notes from Canada's Mining Frontier." *Environmental Justice* 2, no. 3: 117–25.

Kehoe, Myles. 2002. "Military Dumpsites off Canada's Atlantic Coast." Office of the Auditor General of Canada, Petition: No. 50A.

Kendall, Karalyn. 2008. "The Face of a Dog: Levinasian Ethics and Human/Dog Co-evolution." In *Queering the Non/Human*, edited by Noreen Giffney and Myra J. Hird, 185–204. Aldershot, UK: Ashgate Press.

Kennedy, Greg. 2007. *Ontology of Waste: The Disposable and its Problematic Nature*. Albany: State University of New York Press.

Kim, Kwon-Rae, and Gary Owens. 2010. "Potential for Enhanced Phytoremediation of Landfills Using Biosolids: A Review." *Journal of Environmental Management* 91, no. 4: 791–7.

Kimes, Nikole, Amy Callaghan, Joseph Suflita, and Pamela Morris. 2014. "Microbial Transformation of the Deepwater Horizon Oil Spill: Past, Present, and Future Perspectives." *Frontiers in Microbiology* 5, no. 603: 1–11.

King, Gary, Joel Kostka, Terry Hazen, and Patricia Sobecky. 2015. "Deepwater Horizon Oil Spill: From Coastal Wetlands to the Deep Sea." *Annual Review of Marine Science*, 7: 377–401.

King, Wendell C. 2001. "Foreword." In *The Environmental Legacy of Military Operations*, edited by Judy Ehlen and Russell S. Harmon, ix–x. Reviews in Engineering Geology, vol. 14. Boulder, CO: Geological Society of America.

Kingston City Council. 2011. *Report to Environment, Infrastructure and Transportation Policies Committee*. Report no. EITP-11-001. Kingston, ON: City of Kingston.

– 2012. *Environment, Infrastructure and Transportation Policies Committee*. Meeting no. 09-2012. Kingston, ON: City of Kingston. http://www.cityofkingston.ca/documents/10180/984482/EIT_Agenda-0912.pdf/95bd1a0a-6b85-4470-b8d2-36f6210571dc.

– 2013a. *Report to Council: Transfer, Transportation and Disposal of Municipal Solid Waste*. Report no. 13-051. Kingston, ON: City of Kingston. http://www.cityofkingston.ca/documents/10180/58881/COU_A0313-13051.pdf/702016fe-b635-4188-8366-feadc1992b38.

– 2013b. *Report to Environment, Infrastructure and Transportation Policies Committee*. Report no. EITP-13-014. Kingston, ON: City of Kingston. http://www.cityofkingston.ca/documents/10180/1483472/EIT_A0713-13014.pdf/553dab7a-448d-4ef2-a80a-5bfob53cc45d.

– 2013c. *Information Report to the Near Campus Neighbourhoods Advisory Committee*. Report no. NCN-13-004. Kingston, ON: City of Kingston. http://www.cityofkingston.ca/documents/10180/1885123/NCN_A0313-13004.pdf/c989083a-8b10-41db-a984-e7d94a1b7a53.

Kingston Solid Waste Task Force. 2002. *Solid Waste Services Situation Analysis Review*. Kingston, ON: City of Kingston.

*Kingston Whig-Standard*. 1997. "Parties Prepare to Set Trial Date." *Kingston Whig-Standard*, Environmental Bureau of Investigation, 4 June.

– 2009. "Let's Get Serious about Curbing Waste." *Kingston Whig-Standard*, editorial, 21 April.

Kirgin, Harlan. 2011. "Oil-Eating Bacteria Feasted on Oil from Deepwater Horizon's Broken Well Says Scientist." *GulfLive*, 28 July.

Kloppenburg, Jack. 1991. "No Hunting! Biodiversity, Indigenous Rights, and Scientific Poaching." *Cultural Survival Quarterly Magazine* 15, no. 3 (summer 1991): 14–18.

Koerth-Baker, Maggie. 2016. "How Do You Put Out a Subterranean Fire beneath a Mountain of Trash?" *FiveThirtyEight*, 10 May.

Kollikkathara Naushad, Huan Feng, and Eric Stern. 2009. "A Purview of Waste Management Evolution: Special Emphasis on USA." *Waste Management* 29, no. 2: 974–85.

Kornblith, Hilary. 1999. "Knowledge in Humans and Other Animals." *Philosophical Perspectives* 13: 327–46.

Krausz, Robert. 2012. "All for Naught? A Critical Study of Zero Waste to Landfill Initiatives." PhD diss., Lincoln University. http://hdl.handle.net/10182/5301.

Kristeva, Julia. 1982. *Powers of Horror: An Essay on Abjection*. Translated by Leon S. Roudiez. New York: Columbia University Press.

Krupar, Shiloh R. 2013. *Hot Spotter's Report: Military Fables of Toxic Waste*. Minneapolis, MN: University of Minnesota Press.

Kurdve, Martin, Sasha Shahbazi, Marcus Wendin, and Cecilia Bengtsson. 2015. "Waste Flow Mapping to Improve Sustainability of Waste Management: A Case Study Approach." *Journal of Cleaner Production* 98: 304–15.

Kuyek, Joan. 2019. *Unearthing Justice: How to Protect Your Community from the Mining Industry*. Toronto, ON: Between the Lines.

Labban, Mazen. 2014. "Deterritorializing Extraction: Bioaccumulation and the Planetary Mine." *Annals of the Association of American Geographers* 104, no. 3: 560–76.

Lackenbauer, P. Whitney. 2011a. "From Polar Race to Polar Saga: An Integrated Strategy for Canada and the Circumpolar World." In *Canada and the Changing Arctic: Sovereignty, Security, and Stewardship*, edited by Franklyn Griffiths, Rob Huebert, and P. Whitney Lackenbauer, 69–181. Waterloo, ON: Wilfrid Laurier University Press.

– 2011b. "Sovereignty, Security, and Stewardship: An Update." In *Canada and the Changing Arctic: Sovereignty, Security, and Stewardship*, edited by Franklyn Griffiths, Rob Huebert, and P. Whitney Lackenbauer, 227–74. Waterloo, ON: Wilfrid Laurier University Press.

Lackenbauer, P. Whitney, and Matthew Farish. 2007. "The Cold War on

Canadian Soil: Militarizing a Northern Environment." *Environmental History* 12, no. 4: 920–50.

Lackenbauer, P. Whitney, and Ryan Shackleton. 2012. "When the Skies Rained Boxes: The Air Force and the Qikiqtani Inuit, 1941–64." Working Papers on Arctic Security, no. 4. Toronto, ON: Walter and Duncan Gordon Foundation and ArcticNet Arctic Security Projects.

LaDuke, Winona. 2016. *The Winona LaDuke Chronicles: Stories from the Front Lines in the Battle for Environmental Justice*. Halifax, NS: Fernwood Publishing.

La Fay, Howard. 1958. "DEW Line: Sentry of the Far North." *National Geographic* 114, no. 1: 128–46.

Lagus, Todd P. 2005. "Reprocessing of Spent Nuclear Fuel: A Policy Analysis." Minneapolis, MN: University of Minnesota, Washington Internships for Students of Engineering.

Lajeunesse, Adam. 2007. "The Distant Early Warning Line and the Canadian Battle for Public Perception." *Canadian Military Journal* 8, no. 2: 51–9. http://www.journal.forces.gc.ca/vo8/no2/lajeunes-eng.asp.

Land, Nick. 2011. "Barker Speaks: The CCRU Interview with Professor D.C. Barker." In *Fanged Noumena: Collected Writings 1987–2007*, edited by Robin Mackay and Ray Brassier, 493–506. Falmouth, UK: Urbanomic.

Landström, Catharina, Sarah J. Whatmore, Stuart N. Lane, Nicholas A. Odoni, Neil Ward, and Susan Bradley. 2011. "Coproducing Flood Risk Knowledge: Redistributing Expertise in Critical 'Participatory Modelling.'" *Environment and Planning A* 43, no. 7: 1,617–33.

Lane, Stuart N., Nicholas A. Odoni, Catharina Landström, Sarah J. Whatmore, Neil Ward, and Susan Bradley. 2011. "Doing Flood Risk Science Differently: An Experiment in Radical Scientific Method." *Transactions of the Institute of British Geographers* 36, no. 1: 15–36.

Lange, Klas, R. Kerry Rowe, and Heather Jamieson. 2009. "Diffusion of Metals in Geosynthetic Clay Liners." *Geosynthetics International* 16, no. 1: 11–27.

Langston, Nancy. 2003. "Gender Transformed: Endocrine Disruptors in the Environment." In *Seeing Nature through Gender*, edited by Virginia Scharff, 129–66. Lawrence, KS: University of Kansas Press.

– 2010. *Toxic Bodies: Hormone Distruptors and the Legacy of DES*. New Haven, CT: Yale University Press.

LaPensee, Elizabeth W., Traci R. Tuttle, Sejal R. Fox, and Nira Ben-Jonathan. 2009. "Bisphenol A at Low Nanomolar Doses Confers Chemoresistance in Estrogen Receptor-α–Positive and –Negative Breast Cancer Cells." *Environmental Health Perspective* 117, no. 2: 175–80.

Larsen, Anna, Hanna Merrild, and Thomas Christensen. 2009. "Recycling of Glass: Accounting of Greenhouse Gases and Global Warming Contributions." *Waste Management and Research* 27: 754–62.

Lash, Scott, Bronislaw Szerszynski, and Brian Wynne, eds. 1996. *Risk, Environment and Modernity: Towards a New Ecology*. London, UK: Sage Publications.

Lathers, Marie. 2006. "Towards and Excremental Posthumanism: Primatology, Women, and Waste." *Society and Animals* 14, no. 4: 417–36.

Latour, Bruno. 1988. *The Pasteurization of France/Irreductions*. Translated by Alan Sheridan and John Law. Cambridge, MA: Harvard University Press.

– 1992. "Where Are the Missing Masses? The Sociology of a Few Mundane Artifacts." In *Shaping Technology/Building Society: Studies in Sociotechnical Change*, edited by Wiebe E. Biker and John Law, 225–58. Cambridge, MA: MIT Press.

– 1993. *We Have Never Been Modern*. Cambridge, MA: Harvard University Press.

– 1999. *Pandora's Hope: Essays on the Reality of Science Studies*. Cambridge, MA: Harvard University Press.

– 2004. "Why Has Critique Run Out of Steam? From Matters of Fact to Matters of Concern." *Critical Inquiry* 30, no. 2: 225–48.

– 2005. "From Realpolitik to Dingpolitik, or How to Make Things Public." In *Making Things Public: Atmospheres of Democracy*, edited by Bruno Latour and Peter Weibel, 4–31. Cambridge, MA: ZKM Centre for Art and Media, and MIT Press.

– 2007a. *Reassembling the Social: An Introduction to ActorNetwork-Theory*. Oxford, UK: Oxford University Press.

– 2007b. "Turning Around Politics: A Note on Gerard de Vries' Paper." *Social Studies of Science* 37, no. 5: 811–20.

Laugrand, Frederic, and Jarich Oosten. 2010. *Inuit Shamanism and Christianity*. Montreal and Kingston: McGill-Queen's University Press.

Law, John, and John Hassard, eds. 1999. *Actor Network Theory and After*. Oxford, UK: WileyBlackwell.

Lee, Stuart, and Wolff-Michael Roth. 2003. "Science and the 'Good Citizen': Community-Based Scientific Literacy." *Science, Technology and Human Values* 28, no. 3: 403–24.

Lemke, Thomas. 2001. "'The Birth of Bio-Politics': Michel Foucault's Lecture at the Collège de France on Neo-Liberal Governmentality." *Economy and Society* 30, no. 2: 190–207.

Leonard, Annie. 2007. *The Story of Stuff*. Directed by Louis Fox, produced by Erica Priggen. Berkeley, CA: Free Range Studios.

Lepawsky, Josh. 2012. "Legal Geographies of E-Waste Legislation in Canada and the US: Jurisdiction, Responsibility and the Taboo of Production." *Geoforum* 43, no. 6: 1194–206.

– 2018. *Reassembling Rubbish: Worlding Electronic Waste*. Cambridge, MA: MIT Press.

Lepawsky, Josh, and Chris McNabb. 2010. "Mapping International Flows of Electronic Waste." *Canadian Geographer* 54, no. 2: 177–95.

Lepawsky, Josh, and Charles Mather. 2011. "From Beginnings and Endings to Boundaries and Edges: Rethinking Circulation and Exchange through Electronic Waste: From Beginnings and Endings to Boundaries and Edges." *Area* 43, no. 3: 242–9.

LeTourneau, Michele. 2014. "Planning Commission Slams Ottawa." *Northern News Service*, 23 June.

Levinas, Emmanuel. 1988. "Report of 'The Paradox of Morality': An Interview with Emmanuel Levinas." In *The Provocation of Levinas: Rethinking the Other*, edited by Robert Bernasconi and David Wood, 168–80. London, UK: Routledge.

– 1990. "The Name of a Dog, or Natural Rights." In *Difficult Freedom: Essays on Judaism*, translated by Sean Hand, 151–3. Baltimore, MD: Johns Hopkins University Press.

– 2004. "Interview." In *Animal Philosophy: Essential Readings in Continental Thought*, edited by Matthew Calarco and Peter Atterton, 49–50. New York: Continuum.

Levis, James, Morton Barlaz, Nickolas Themelis, and Priscilla Ulloa. 2010. "Assessment of the State of Food Waste Treatment in the United States and Canada." *Waste Management* 30, no. 8–9: 1486–94.

Li, Tania. 2007. *The Will to Improve: Governmentality, Development, and the Practice of Politics*. Durham, NC: Duke University Press.

Li, Allen Lunzhu, and R. Kelly Rowe. 2001. "Influence of Creep and Stress Relaxation of Geosynthetic Reinforcement on Embankment Behaviour." *Geosynthetics International* 8, no. 3: 233–70.

Liboiron, Max. 2010. "Recycling as a Crisis of Meaning." *Topia: Canadian Journal of Cultural Studies* 4: 1–9.

– 2013. "Modern Waste as Strategy." *Lo Squaderno: Explorations in Space and Society*, 29: 9–12.

– 2018 "Waste Colonialism." *Discard Studies*, 1 November. https://discardstudies.com/2018/11/01/waste-colonialism.

Liboiron, Max, Manuel Tironi, and Nerea Calvillo. 2018. "Toxic Politics:

Acting in a Permanently Polluted World." *Social Studies of Science* 48, no. 3: 331–49.

Lippmann, Walter. 1993. *The Phantom Public.* Introduction by Wilfred M. McClay. New Brunswick, NJ: Transaction Publishers.

Livingstone, Andrew 2013. "One Giant Mess: Feds Told to Find Permanent Solution for Giant Mine Cleanup." CIM *Magazine* (September): 28–9.

Ljunggren, David. 2009. "Every G20 Nation Wants to Be Canada, Insists PM." Reuters, 25 September.

Llewelyn, John. 1991. "Am I Obsessed by Bobby? (Humanism of the Other Animal)." In *Re-Reading Levinas*, edited by Robert Bernasconi and Simon Critchley, 234–46. Bloomington: Indiana University Press.

Locke, John. 2011. *Second Treatise of Government*, edited by Harold Laski. New York: CreateSpace Independent Publishing Platform.

Longino, Helen E. 1990. *Science as Social Knowledge: Values and Objectivity in Scientific Inquiry*. Princeton, NJ: Princeton University Press.

Loock, D. 2014. "Distant Early Warning (DEW) Line." In *Antarctica and the Arctic Circle: A Geographic Encyclopedia of the Earth's Polar Regions*, vol. 1: A–I, edited by Andrew J. Hund, 229–31. Santa Barbara, CA: ABC-CLIO Greenwood.

Lottermoser, Bernd G. 2010. *Mine Wastes: Characterization, Treatment, and Environmental Impacts*, 3rd edition. Berlin, Germany: Springer-Verlag.

Lougheed, Scott. 2017. "Disposing of Risk: The Biopolitics of Recalled Food and the (Un)Making of Waste." PhD thesis, Queen's University.

Lougheed, Scott Cameron, Myra J. Hird, and R. Kerry Rowe. 2016. "Governing Household Waste Management: An Empirical Analysis and Critique." *Environmental Values* 25, no. 3: 287–308.

Ludwig, Christian, Stefanie Hellweg, and Samuel Stucki. 2003. "The Problem with Waste." In *Municipal Solid Waste Management: Strategies and Technologies for Sustainable Solutions*, edited by Christian Ludwig, Stefanie Hellweg, and Samuel Stucki. Berlin, Germany: Springer-Verlag.

Luton Larry S. 1996. *The Politics of Garbage: A Community Perspective on Solid Waste Policy Making*. Pittsburgh, PA: University of Pittsburgh Press.

Lynas Mark. 2012. *The God Species: How Humans Really Can Save the Planet*. London, UK: Fourth Estate.

Lyotard, Jean-François. 1989. The *Differend: Phrases in Dispute.*

Translated by Georges Van Den Abbeele. Minneapolis, MN: University of Minnesota Press.

– 1991. *The Inhuman.* Translated by Geoffrey Bennington and Rachel Bowlby. Stanford, CA: Stanford University Press.

MacAlpine, Ian. 2012. "Most People Get One-Bag Policy." *Kingston Whig-Standard*, 10 September.

MacAlpine, Ian. 2015. "Former Owner of Scott Environmental Charged in Relation to Storage of Hazardous Waste." *Kingston Whig-Standard*, 28 July.

MacBride, Samantha. 2012. *Recycling Reconsidered: The Present Failure and Future Promise of Environmental Action in the United States.* Cambridge, MA: MIT Press.

MacFarlane, Allison. 2003. "Underlying Yucca Mountain: The Interplay of Geology and Policy in Nuclear Waste Disposal." *Social Studies of Science* 33, no. 5: 783–807.

Mackey, Eva. 2016. *Unsettled Expectations: Uncertainty, Land and Settler Decolonization.* Halifax, NS: Fernwood Publishing.

Maclaren, Virginia W. 2010a. "Urban Waste Management." In *Encyclopedia of Geography*, edited by Barney Warf, 2978–81. Los Angeles, CA: Sage Publications.

– 2010b. Waste Management: Moving Up the Hierarchy. In *Resource and Environmental Management: Addressing Conflict and Uncertainty*, 4th ed., edited by Bruce Mitchell, 424–52. Toronto, ON: Oxford University Press.

Maclaren, Virginia W., and Nguyen Anh Thi Thu, eds. 2003. *Gender and Waste Management: Vietnamese and International Experiences.* Hanoi, Vietnam: National Political Publisher.

Mahoney, Nick, Janet Newman, and Clive Barnett. 2010. "Introduction: Rethinking the Public." In *Rethinking the Public: Innovations in Research, Theory and Politics*, edited by Nick Mahoney, Janet Newman, and Clive Barnett. Bristol, UK: Policy Press.

Makhijani, Arjun, and Stephen I. Schwartz. 1998. "Victims of the Bomb." In *Atomic Audit: The Costs and Consequences of US Nuclear Weapons Since 1940*, edited by Stephen I. Schwartz, 395–432. Washington, DC: Brookings Institute.

Makivik Corporation. 2009. *Regarding the Slaughtering of Nunavik "Qimmiit" (Inuit Dogs) from the Mid-1950s to the Late 1960s.* Quebec City, QC: Minister of Indian and Northern Affairs for the Government of Canada.

Malamud, Randy. 2013. "Foreword." In *Trash Animals: How We Live with Nature's Filthy, Feral, Invasive, and Unwanted Species*, edited by Kelsey Nagy and Phillip David Johnson II, 11–13. Minneapolis, MN: University of Minnesota Press.

Malpas, Jeff, and Gary Wickham. 1995. "Governance and Failure: On the Limits of Sociology." *Journal of Sociology* 31, no. 3: 37–50.

Maniates, Michael. 2002. "Individualization: Plant a Tree, Buy a Bike, Save the World?" In *Confronting Consumption*, edited by Thomas Princen, Michael Maniates, and Ken Conca, 43–66. Cambridge, MA: MIT Press.

Marcus, Alan R. 1992. *Out in the Cold: The Legacy of Canada's Inuit Relocation Experiment in the High Arctic*. IWGIA Document 71. Copenhagen, Denmark: International Working Group for Indigenous Affairs.

Margulis, Lynn. 1998. *Symbiotic Planet: A New Look at Evolution*. London, UK: Phoenix.

Marres, Noortje S. 2005a. "No Issue, No Public: Democratic Deficits after the Displacement of Politics." PhD diss., University of Amsterdam. https://pure.uva.nl/ws/files/3890776/38026_thesis_nm_final.pdf.

– 2005b. "Issues Spark a Public into Being." In *Making Things Public: Atmospheres of Democracy*, edited by Bruno Latour and Peter Weibel, 208–17. Karlsruhe and Cambridge, MA: ZKM Centre for Art and Media and MIT Press.

– 2007. "The Issues Deserve More Credit: Pragmatist Contributions to the Study of Public Involvement in Controversy." *Social Studies of Science* 37, no. 5: 759–80.

Masco, Joseph. 2004. "Mutant Ecologies: Radioactive Life in Post–Cold War New Mexico." *Cultural Anthropology* 19, no. 4: 517–50.

– 2006. *The Nuclear Borderlands: The Manhattan Project in Post-Cold War New Mexico*. Princeton, NJ: Princeton University Press.

– 2013. "Planetary Optics: The Age of Fallout." Paper presented at the Constructions of Globality: Ideas, Image and Artefacts Conference, 18 June, Copenhagen, Denmark.

Mbembe, Achille. 2003. "Necropolitics." Translated by Libby Meintjes. *Public Culture* 15, no. 1: 11–40.

McCluskey, Kerry. 2013. *Tulugaq: An Oral History of Ravens*. Iqaluit, NU: Inhabit Media.

McGovern, Dan. 1995. *The Campo Indian Landfill War: The Fight for Gold in California's Garbage*. Norman, OK: University of Oklahoma Press.

McGrath, Janet Tamalik. 2012. "Isumaksaqsiurutigijakka: Conversations with Aupilaarjuk Towards a Theory of Inuktitut Knowledge Renewal." PhD diss., Carleton University.

McGregor, Deborah. 2009. "Linking Traditional Knowledge and Environmental Practice in Ontario" *Journal of Canadian Studies* 43, no. 3: 69–100.

McHugh, Susan. 2013. "'A Flash Point in Inuit Memories': Endangered Knowledges in the Mountie Sled Dog Massacre." *ESC: English Studies in Canada* 39, no. 1: 149–75.

McKenzie-Mohr, Doug. 2000. "New Ways to Promote Proenvironmental Behavior: Promoting Sustainable Behavior: An Introduction to Community-Based Social Marketing." *Journal of Social Issues* 56, no. 3: 543–54.

McKenzie-Mohr, Doug, Lisa Sara Nemiroff, Laurie Beers, and Serge Desmarais. 1995. "Determinants of Responsible Environmental Behavior." *Journal of Social Issues* 51, no. 4: 139–56.

McKerlie, Kate, Nancy Knight, and Beverley Thorpe. 2006. "Advancing Extended Producer Responsibility in Canada." *Journal of Cleaner Production* 14, no. 6–7: 616–28.

McMahon, Kevin. 1988. *Arctic Twilight: Reflections on the Future of Canada's Northern Land and People*. Toronto, ON: James Lorimer and Company.

McRobert, David. 1994. "Ontario's Blue Box System: A Case Study of Government's Role in the Technological Change Process, 1970–1991." LLM thesis, Osgoode Hall Law School, York University.

Meillassoux, Quentin. 2008. *After Finitude: An Essay on the Necessity of Contingency*. New York: Continuum.

Melosi, Martin V. 2005. *Garbage in the Cities: Refuse, Reform, and the Environment*. Revised edition. Pittsburgh, PA: University of Pittsburgh Press.

Metro Vancouver. 2007. "Burns Bog Ecological Conservancy Management Plan." Vancouver, BC: Metro Vancouver. http://www.metrovancouver.org/services/parks/ParksPublications/BurnsBogManagementPlan.pdf.

– 2010. "Integrated Solid Waste and Resource Management: A Solid Waste Management Plan." Vancouver, BC: Metro Vancouver. http://www.metrovancouver.org/services/solid-waste/SolidWastePublications/ISWRMP.pdf.

– 2013. "Recycling and Solid Waste Management, 2013 Report." Vancouver, BC: Metro Vancouver. http://www.metrovancouver.org/services/solid-waste/SolidWastePublications/2013_Solid_Waste_Management_Annual_Summary.pdf

– 2014a. "Bylaw 280." Vancouver, BC: Metro Vancouver.
– 2014b. "Metro Vancouver Budget in Brief." Vancouver, BC: Metro Vancouver.
Metuzals, Jessica, and Myra J. Hird. 2018. "'The Disease that Knowledge Must Cure': Sites of Uncertainty in Arctic Development." In *Arctic Yearbook: In Theory and Practice*, edited by Lassi Heininen and Heather Exner-Pirot. Akureyri, Iceland: Northern Research Forum.
Michael, Mike. 1996. "Ignoring Science: Discourses of Ignorance in the Public Understanding of Science." In *Misunderstanding Science? The Public Reconstruction of Science and Technology*, edited by Alan Irwin and Brian Wynne, 107–25. Cambridge, MA: Cambridge University Press.
Millar, Ray. 2008. "Letter to Minister of the Environment John Gerretsen RE: Proposed County of Simcoe Landfill Site 41." Stop Dump Site 41.
Miller, Jon D. 1991. *The Public Understanding of Science and Technology in the United States, 1990: A Report to the National Science Foundation*. DeKalb, IL: Public Opinion Laboratory, Northern Illinois University.
– 2004. "Public Understanding of, and Attitudes toward, Scientific Research: What We Know and What We Need to Know." *Public Understanding of Science* 13: 273–94.
Milloy, John S. 1999. *A National Crime: The Canadian Government and the Residential School System, 1879 to 1986*. Winnipeg, MB: University of Manitoba Press.
Ministry of Environment (Ontario). 2010. *Annual Report*. Chapter 4, section 4.09. "Non-Hazardous Waste Disposal and Diversion." Toronto, ON: Government of Ontario.
Ministry of Environment and Climate Change (MOECC). 2015. "Court Bulletin: Waste Transfer Business and Former Owner Fined $130,000 Total for Non-Compliance with Ministry Approval." Toronto, ON: Government of Ontario. https://news.ontario.ca/ene/en/2015/07/waste-transfer-business-and-former-owner-fined-130000-total-for-non-compliance-with-a-ministry-appro.html.
Mitchell, Marybelle. 1996. *From Talking Chiefs to a Native Corporate Elite*. Montreal and Kingston: McGill-Queen's University Press.
Mitchell, Timothy. 2002. "Can the Mosquito Speak?" In *Rule of Experts: Egypt, Techno-Power, Modernity*, edited by Thomas Nagel, 19–53. Los Angeles, CA: University of California Press.
Monbiot, George. 2019 "We Cannot Change the World by Changing Our Buying Habits." *Guardian*, 6 November.

Monastersky, Richard. 2015. "First Atomic Blast Proposed as Start of Anthropocene." *Nature*, 16 January.

Monstadt, Jochen. 2009. "Conceptualizing the Political Ecology of Urban Infrastructures: Insights from Technology and Urban Studies." *Environment and Planning A* 41, no. 8: 1924–42.

Moore, Sarah A. 2012. "Garbage Matters: Concepts in New Geographies of Waste." *Progress in Human Geography* 36, no. 6: 780–99.

Mothiba, Mathema, Shadung Moja, and Chris Loans. 2017. "A Review of the Working Conditions and Health Status of Waste Pickers at Some Landfill Sites in the City of Tshwane Metropolitan Municipality, South Africa." *Advances in Applied Science Research* 8, no. 3: 90–7.

Murphy, David. 2014a. "Iqaluit Residents with Bad Lungs Should Stay Indoors, Expert Says." *Nunatsiaq News*, 22 May.

– 2014b. "Landfill Expert Says $3.5-Million Dunking Best Solution for Iqaluit Dump Fire." *Nunatsiaq News*, 1 July.

– 2014c. "Pregnant Nunavut Mom Worried about Dump Smoke Toxins." *Nunatsiaq News*, 4 July.

Mydans, Carl. 1963. "Flesh Freezes Solid in 30 Seconds." *Life Magazine*, 1 March, 27.

Myers, Heather. 2001. "Changing Environment, Changing Times: Environmental Issues and Political Action in the Canadian North." *Environment: Science and Policy for Sustainable Development* 43, no. 6: 32–44.

Myers, Heather, and Don Munton. 2000. "Cold War, Frozen Wastes: Cleaning Up the DEW Line." *Environment and Sécurité* 4: 119–38.

Nadasdy, Paul. 1999. "The Politics of TEK: Power and the 'Integration' of Knowledge." *Arctic Anthropology* 36, no. 1–2: 1–18.

– 2005. "The Anti-Politics of TEK: The Institutionalization of Co-Management Discourse and Practice." *Anthopologica* 47, no. 2: 215–32.

Nagel, Jeff. 2015. "Metro Vancouver Halts Plan to Build New Garbage Incinerator." *The Now*, 10 December.

Nagel, Thomas. 1974. "What Is It like to Be a Bat?" *Philosophical Review* 83, no. 4: 435–50.

Nagy, Kelsey, and Phillip David Johnson II. 2013. "Introduction." In *Trash Animals: How We Live with Nature's Filthy, Feral, Invasive, and Unwanted Species*, edited by Kelsey Nagy and Phillip David Johnson II, 1–27. Minneapolis: University of Minnesota Press.

National Energy Board. 2012. *June 2012 Community Engagement Report*. Ottawa, ON: National Energy Board.

National Defence. 2014. "Operation NANOOK." Ottawa, ON: Government of Canada, National Defence.

Ni, Hong-Gang, Hui Zeng, Shu Tao, and Eddy Y. Zeng. 2010. "Environmental and Human Exposure to Persistent Halogenated Compounds Derived from E-Waste in China." *Environmental Toxicology and Chemistry* 29, no. 6: 1237–47.

Nickels, Scott, Karen Kelley, Carrie Grable, Martin Lougheed, and James Kuptana. 2012. *Nilliajut: Inuit Perspectives on Security, Patriotism and Sovereignty*. Ottawa, ON: Inuit Tapiriit Kanatami.

Nixon, Rob. 2011. *Slow Violence and the Environmentalism of the Poor*. Cambridge, MA: Harvard University Press.

Norris, Mike. 2011. "Two Bags? Too Many." *Kingston Whig-Standard*, 2 November.

Northern Strategy. 2015. "Exercising Our Arctic Sovereignty." Ottawa, ON: Government of Canada, Canadian Northern Economic Development Agency.

Nuclear Energy Agency. 2010. *Radioactive Waste in Perspective*. Paris, France: Organisation for Economic Cooperation and Development.

*Nunatsiaq News*. 2014. "Parks Day in Nunavut's Capital Cancelled due to Dump Smoke." *Nunatsiaq News*, 17 July.

Nunavut Department of Environment. N.d. "Solid Waste Management in Nunavut: A Backgrounder." Iqaluit, NU: Government of Nunavut.

– 2010. "Guideline for the General Management of Hazardous Waste." Iqaluit, NU: Government of Nunavut.

Nungak, Zebedee. 2004. "Ratcheting Garbage to a Federal Affair?" *Windspeaker Publication* 2, no. 8: 18–22.

– 2006. "NASIVVIK: Introducing the Science of Qallunology." *Windspeaker Publication* 24, no. 2: 18.

O'Reilly, Kevin. 2013. "The Giant Mine: Will Reason Prevail?" *Alternatives North*, 25 October. https://miningwatch.ca/blog/2013/10/25/giant-mine-will-reason-prevail.

Obbard, Rachel W., Saeed Sadri, Ying Qi Wong, Alexandra A. Khitun, Ian Baker, and Richard C. Thompson. 2014. "Global Warming Releases Microplastic Legacy Frozen in Arctic Sea Ice." *Earth's Future* 2, no. 6: 315–20.

Oels, Angela. 2006. "Rendering Climate Change Governable: From Biopower to Advanced Liberal Government?" *Journal of Environmental Policy and Planning*, 7: 185–207.

Office of the Prime Minister. 2013. "PM Harper Delivers Remarks in Hay River, Northwest Territories." Ottawa, ON: Office of the Prime Minister.

– 2014. "PM Harper Arrives in Iqaluit for the Last Stop of his Ninth Annual Northern Tour." Ottawa, ON: Office of the Prime Minister.
Olmer, Naya, Bryan Comer, Biswayjoy Roy, Xiaoli Mao, and Dan Rutherford. 2017. *Greenhouse Gas Emissions from Global Shipping, 2013–2015*. Washington, DC: International Council on Clean Transportation.
Olson, Philip, R. 2015. "Knowing 'Necro-Waste': A Reply to Hird." *Social Epistemology: A Journal of Knowledge, Culture and Policy* 2, no. 7: 1–20.
Ongondo, Francis O., Ian D. Williams, and Tom J. Cherrett. 2011. "How Are WEEE Doing? A Global Review of the Management of Electrical and Electronic Wastes." *Waste Management* 31, no. 4: 714–30.
Ontario Executive Council. 1996. *Order in Council 516/96*. Toronto, ON: Government of Ontario, Ontario Executive Council. http://www.ert.gov.on.ca/files/DEC/8703d2.pdf.
Ontario Ministry of the Environment. 1998. *Provisional Certificate of Approval for a Waste Disposal Site*. Toronto, ON: Government of Ontario.
Ottinger, Gwen. 2010. "Buckets of Resistance: Standards and the Effectiveness of Citizen Science" *Science, Technology, and Human Values* 35, no. 2: 244–70.
Ontario Waste Management Association (OWMA). 2016. "State of Waste in Ontario: Landfil Report." 1st Annual Landfill Report. Brampton, ON: Ontario Waste Management Association.
Packard, Vance. 1960. *The Waste Makers*. New York: Ig Publishing.
Paine, Robert, ed. 1977. *The White Arctic: Anthropological Essays on Tutelage and Ethnicity*. St John's, NL: Memorial University of Newfoundland.
Parizeau, Kate. 2006. "A World of Trash: From Canada to Cambodia, Waste Is a Common Problem with Common Solutions." *Alternatives Journal* 32, no. 1: 16–18.
– 2016. "Witnessing Urban Change: Insights from Informal Recyclers in Vancouver, BC." *Urban Studies* 54, no. 8: 1921–37.
Parizeau, Kate, and Josh Lepawsky. 2015. "Legal Orderings of Waste in Built Spaces." *International Journal of Law in the Built Environment* 7, no. 1: 21–38.
Parliament of Canada. 1995. "Evidence – May 9, 1995." Ottawa, ON: Parliament of Canada, House of Commons. http://www.parl.gc.ca/content/hoc/archives/committee/351/sust/evidence/121_95-05-09/sust121_blk-e.html.

- 2003. "The Federal Role in Waste Management." Ottawa, ON: Library of Parliament.
Parrilla, Leslie. 2012. "Falcons Protect Landfill from Trash-Stealing Gulls." *Press Enterprise.* 18 November.
Parsons, Liz. 2008. "Thompsons' Rubbish Theory: Exploring the Practices of Value Creation." *European Advances in Consumer Research* 8: 390–3.
Patriquin, Martin. 2015. "Move Over, Mushers: A Battle Brews Over Sled Dog Parking in Iqaluit." *Maclean's.* 20 January.
Paudyn, Krysta, Allison Rutter, R. Kerry Rowe, and John S. Poland. 2008. "Remediation of Hydrocarbon Contaminated Soils in the Canadian Arctic by Landfarming." *Cold Regions Science and Technology* 53, no. 1: 102–14.
Pearce, Tristan D., James D. Ford, Jason Prno, Frank Duerden, Jeremy Pittman, Maude Beaumier, Lea Berrang-Ford, and Barry Smit. 2010. "Climate Change and Mining in Canada." *Mitigation Adaptation Strategies for Global Change* 16, no. 3: 347–68.
Perić, Sabrina. 2015. "Darwin's North: Military Adaptation and the Birth of Counterinsurgency in the Land of Oil and Ice." Paper presented at the American Anthropological Association, 2015 Annual Meeting, 21 November.
Petroski, Henry. 2001. "The Success of Failure." *Technology and Culture* 42, no. 2: 321–8.
Petts, Judith. 1997. "The Public-Expert Interface in Local Waste Management Decisions: Expertise, Credibility and Process." *Public Understanding of Science* 6: 359–81.
– 1998. "Trust and Waste Management Information Expectation Versus Observation." *Journal of Risk Research* 1, no. 4: 307–20.
– 2001. "Evaluating the Effectiveness of Deliberative Processes: Waste Management Case Studies." *Journal of Environmental Planning and Management* 44, no. 2: 207–26.
Pickard, William F. 2010. "Finessing the Fuel: Revisiting the Challenge of Radioactive Waste Disposal." *Energy Policy* 38, no. 2: 709–14.
Pickering, Andrew. 1995. *The Mangle of Practice: Time, Agency, and Science.* Chicago, IL: University of Chicago Press.
Pielke Jr., Roger A. 2004. "When Scientists Politicize Science: Making Sense of Controversy over the Skeptical Environmentalist." *Environmental Science and Policy* 7, no. 5: 405–17.
– 2007. *The Honest Broker: Making Sense of Science in Policy and Politics.* Cambridge, MA: Cambridge University Press.

Pier, M. Dawn, Alexandra A. Betts-Piper, Christopher C. Knowlton, Barbara A. Zeeb, and Kenneth J. Reimer. 2003. "Redistribution of Polychlorinated Biphenyls from a Local Point Source: Terrestrial Soil, Freshwater Sediment, and Vascular Plants as Indicators of the Halo Effect." *Arctic, Antarctic, and Alpine Research* 35, no. 3: 349–60.

Pigott, Peter. 2011. *From Far and Wide: A Complete History of Canada's Arctic Sovereignty*. Toronto, ON: Dundurn.

Piper, Liza. 2009. *The Industrial Transformation of Subarctic Canada*. Vancouver, BC: University of British Columbia Press.

Pizzolato, Larissa, Stephen E.L. Howell, Chris Derksen, Jackie Dawson, and Luke Copland. 2014. "Changing Sea Ice Conditions and Marine Transportation Activity in Canadian Arctic Waters Between 1990 and 2012." *Climate Change* 123, no. 1: 161–73.

PlasticsEurope. n.d. "Association of Plastics Manufacturers." PlasticsEurope. http://www.plasticseurope.org.

Plumwood, Val. 1993. *Feminism and the Mastery of Nature*. London and New York: Routledge.

Poland, John S., Scott Mitchell, and Allison Rutter. 2001. "Remediation of Former Military Bases in the Canadian Arctic." *Cold Regions Science and Technology* 32, no. 2–3: 93–105.

Polar Life. n.d. "Tulugaq – Raven." Polar Life.

Pollans, Lily B. 2017. "Trapped in Trash: 'Modes of Governing' and Barriers to Transitioning to Sustainable Waste Management." *Environment and Planning A* 49, no. 10: 2300–23.

Popli, Rakesh. 1999. "Scientific Literacy for All Citizens: Different Concepts and Contents," *Public Understanding of Science* 8, no. 2: 123–37.

Powell, Maria, Sharon Dunwoody, Robert Griffin, and Kurt Neuwirth. 2007. "Exploring Lay Uncertainty about an Environmental Health Risk." *Public Understanding of Science* 16, no. 3: 323–43.

Prakash, Nidhi. 2015. "Canada Is Dumping Tons of Garbage in the Philippines." *Splinter*, 14 July.

Price, Jackie. 2007. "Tukisivallialiqtakka: The Things I Have Now Begun to Understand: Inuit Governance, Nunavut and the Kitchen Consultation Model." MA thesis, University of Victoria.

– 2013. "IGOV Indigenous Speaker Series – Jackie Price 'But You're Inuk, Right?'" Public lecture for Indigenous Governance Program, University of Victoria, BC. YouTube video, 1:31:44. 8 April. https://www.youtube.com/watch?v=W36cGxXpjWw.

Public Safety Canada. 2015. "During an Emergency." Ottawa, ON: Get Preparcd. http://www.getprepared.gc.ca/cnt/hzd/drng-en.aspx.

Public Works Canada. 1992. *Literature Review on Abandoned and Waste Disposal Sites in Iqaluit Area, Northwest Territories*. Edmonton, AB: Environmental Services, Pacific Western Region.

PWC. 1992. "Literature Review on Abandoned and Waste Disposal Sites in Iqaluit Area, Northwest Territories." Edmonton, AB: Environmental Services, Pacific Western Region.

Qikiqtani Inuit Association (QIA). 2010. *Qikiqtani Truth Commission Final Report:*

*Achieving Saimaqtigiiniq, 2010*. Iqaluit, NU: Qikiqtani Inuit Association. http://www.qtcommission.com.

– 2013a. "Inuit Sled Dogs in Baffin Region." Iqaluit, NU: Qikiqtani Truth Commission.

– 2013b. *QTC Final Report: Achieving Saimaqatiqiingniq – Thematic Reports and Special Studies, 1950–1975*. Qikiqtani Truth Commission. Iqaluit, NU: Inhabit Media.

– 2013c. *Nuutauniq: Moves in Inuit Life – Thematic Reports and Special Studies, 1950–1975*. Qikiqtani Truth Commission. Iqaluit, NU: Inhabit Media.

– 2013d. *Qimmiliriniq: Inuit Sled Dogs in Qikiqtaaluk – Thematic Reports and Special Studies, 1950–1975*. Qikiqtani Truth Commission. Iqaluit, NU: Inhabit Media.

– 2013e. *Analysis of the RCMP Sled Dog Report – Thematic Reports and Special Studies, 1950–1975*. Qikiqtani Truth Commission. Iqaluit, NU: Inhabit Media.

– 2013f. *The Official Mind of Canadian Colonialism – Thematic Reports and Special Studies, 1950–1975*. Qikiqtani Truth Commission. Iqaluit, NU: Inhabit Media.

Qikiqtani Truth Commission (QTC). *See* Qikiqtani Inuit Association (QIA).

Qitsualik-Tinsley, Rachel. 2013. "Inummarik: Self-Sovereignty in Classic Inuit Thought." In *Nilliajut: Inuit Perspectives on Security, Patriotism, and Sovereignty*, edited by Scott Nickels, Karen Kelley, Carrie Grable, Martin Lougheed, and James Kuptana, 23–34. Ottawa, ON: Inuit Tapiriit Kanatami.

Qitsualik-Tinsley, Rachel, and Sean Qitsualik-Tinsley. 2015. *How Things Came to Be: Inuit Stories of Creation*. Iqaluit, NU: Inhabit Media.

Rabe, Barry G. 1992. "When Siting Works, CanadaStyle." *Journal of Health Politics, Policy, and Law* 17, no. 1: 119–42.

Rabson, Mia. 2019. "Canada Hasn't Issued Any Permits for Companies to Ship Waste, Government Says." CBC News, 29 May.

Rathje, William L., and Cullen Murphy. 2001. *Rubbish! An Archaeology of Garbage*. Tucson: University of Arizona Press.

Reimer, Kenneth J., et al. 1993. *Environmental Study of Eleven DEW Line Sites*. Ottawa, ON: Department of National Defence.

Rennie, Steve. 2014a. "Nutrition North Food Subsidy Program: What Went Wrong." Canadian Press, CBC News, 21 December.

– 2014b. "Military Fretted Over Fumes from Iqaluit's 'Dumpcano.'" Canadian Press, CBC News, 29 October.

Reno, Joshua. 2011. "Beyond Risk: Emplacement and the Production of Environmental Evidence." *American Ethnologist* 38, no. 3: 516–30.

– 2016. *Waste Away: Working and Living with a North American Landfill*. Oakland, CA: University of California Press.

– 2020. *Military Waste: The Unexpected Consequences of Permanent War Readiness*. Oakland, CA: University of California Press.

Revkin, Andrew C. 2012. "How Rachel Carson Spurred Chemical Concerns by Highlighting Uncertainty." *New York Times*, 17 September.

Ristau, Carolyn A. 1991. "Aspects of the Cognitive Ethology of an Injury-Feigning Bird, the Piping Plover." In *Cognitive Ethology: The Minds of Other Animals*, edited by Carolyn A. Ristau, 91–126. New York: Lawrence Erlbaum Associates.

Rittel, Horst, and Melvin Webber. 1973. "Dilemmas in a General Theory of Planning." *Policy Sciences*, 4: 155–69.

Roberts, Celia. 2007. *Messengers of Sex: Hormones, Biomedicine and Feminism*. Cambridge, MA: Cambridge University Press.

Rogers, Heather. 2005. *Gone Tomorrow: The Hidden Life of Garbage*. New York: New Press.

Rogers, Sarah. 2015. "QIA 'Whole-Heartedly' Supports Clyde River Appeal." *Nunatsiaq News*, 23 April.

Rohner, Thomas. 2014. "Nunavut's Most Notorious Dump Fire Could Be Put Out Soon." *Nunatsiaq News*, 26 August.

Rose, Deborah Bird. 2003. "Decolonizing the Discourse of Environmental Knowledge in Settler Societies." In *Culture and Waste: The Creation and Destruction of Value*, edited by Gay Hawkins and Stephen Muecke, 53–72. London, UK: Rowman and Littlefield Publishers.

Rosenfeld, Jonathan. 2013. "Commercial Truck Fatality Statistics." Chicago, IL: Rosenfeld Injury Lawyers.

Roth, Wolff-Michael, and Jacques Désautels. 2004. "Educating for Citizenship: Reappraising the Role of Science Education." *Canadian Journal of Science, Mathematics and Technology Education* 4, no. 2: 149–68.

Rouse, Joseph. 2004. "Barad's Feminist Naturalism." *Hypatia* 19, no. 1: 142–61.
Rowe, Gene, Dee Rawsthorne, Tracey Scarpello, and Jack R. Dainty. 2010. "Public Engagement in Research Funding: A Study of Public Capabilities and Engagement Methodology." *Public Understanding of Science* 19, no. 2: 225–39.
Rowe, R. Kerry. 1988. "Contaminant Migration through Groundwater: The Role of Modeling in the Design of Barriers." *Canadian Geotechnical Journal* 25, no. 4: 778–98.
– 1991. "Contaminant Impact Assessment and the Contaminating Lifespan of Landfills." *Canadian Journal of Civil Engineering* 18, no. 2: 244–53.
– 2004. "Review of Responses re: Final Design for Site 41, County of Simcoe."
– 2005. "Long-Term Performance of Contaminant Barrier Systems, 45th Rankine Lecture." *Géotechnique* 55, no. 9: 631–78.
– 2007. "Advances and Remaining Challenges for Geosynthetics in Geoenvironmental Engineering Applications, 23rd Manual Rocha Lecture." *Soils and Rocks* 30: 3–30.
– 2012. "Design and Construction of Barrier Systems to Minimize Environmental Impacts Due to Municipal Solid Waste Leachate and Gas." Third Indian Geotechnical Society, Ferroco Terzaghi Oration. *Indian Geotech Journal* 42, no. 4: 223–56.
– Rowe, R. Kerry, ed. 2001. *Geotechnical and Geoenvironmental Engineering Handbook*. Norwell, MA: Kluwer Academic Publishing.
Rowe, R. Kerry, Azadeh Hoor, and Andrew Pollard. 2010. "Numerical Examination of a Method for Reducing the Temperature of Municipal Solid Waste Landfill Liners." ASCE *Journal of Environmental Engineering* 136, no. 8: 794–803.
Rowe, R. Kerry, Laura Bostwick, and Richard Thiel. 2010. "Shrinkage Characteristics of Heattacked GCL Seams." *Geotextiles and Geomembranes* 28, no. 4: 352–9.
Rowe, R. Kerry, and Mohammad Zahidul Islam. 2009. "Impact on Landfill Liner Time: Temperature History on the Service Life of HDPE Geomembranes" *Waste Management* 29, no. 10: 2,689–99.
Rowe, R. Kerry, Robert M. Quigley, Richard W.I. Brachman, and John R. Booker. 2004. *Barrier Systems for Waste Disposal Facilities*, 2nd edition. New York: Spon Press.
Rowell, Thelma. 1991. "Till Death Do Us Part: Long-Lasting Bonds between Ewes and Their Daughters." *Animal Behavior* 42, no. 4: 681–2.

– 1993. "Reification of Social Systems." *Evolutionary Anthropology* 2, no. 4: 135–7.
Royal Canadian Mounted Police (RCMP). 2006. *Final Report: RCMP Review of Allegations Concerning Inuit Sled Dogs*. Ottawa, ON: Community, Contract and Aboriginal Policing Services.
Ruff, Kathleen. 2019. "Canadian Government Challenged to Support UN Ban on Exporting Wastes to Developing Countries." RightOnCanada. 21 July. http://www.RightOnCanada.ca.
Rumsfeld, Donald. 2002. "Donald Rumsfeld Unknown Unkowns!" US Department Defense Briefing. YouTube video, 0:35. http://www.youtube.com/watch?v=GiPe1OiKQuk.
Russell, Paul. 2013. "Todays Letters: Ideas for Solving the 'Native Issue.'" *National Post*, 14 January.
Rutherford, Stephanie. 2007. "Green Governmentality: Insights and Opportunities in the Study of Nature's Rule." *Progress in Human Geography* 31, no. 3: 291–307.
Sager, Josh. 2014. "Fracking Floods the Earth with Biocides" *Progressive Cynic*, 19 May. http://theprogressivecynic.com/2014/05/19/fracking-floods-the-earth-with-biocides.
Sakiagaq, Papikattuq. 2009. "Regarding the Slaughtering of Nunavik 'Qimmiit' (Inuit Dogs) from the Mid-1950s to the Late 1960s." In *Makivik Corporation for Minister of Indian and Northern Affairs for the Government of Canada*, 11.
Sandlos, John, and Arn Keeling. 2012. "Claiming the New North: Development and Colonialism at the Pine Point Mine, Northwest Territories, Canada." *Environment and History* 18, no. 1: 5–34.
Sandlos, John, and Arn Keeling. 2015. "Aboriginal Communities, Traditional Knowledge, and the Environmental Legacies of Extractive Development in Canada." *Extractive Industries and Society* 3, no. 2: 278–87.
– 2017. "The Giant Mine's Long Shadow: Arsenic Pollution and Native Peoples in Yellowknife, Northwest Territories." In *Mining North America: An Environmental History since 1522*, edited by John R. McNeill and George Vrtis, 280–312. Los Angeles, CA: University of California Press.
Saul, John Ralston. 2008. *A Fair Country: Telling Truths About Canada*. Toronto, ON: Penguin Canada.
– 2017. *Reimagines Canada*. New York: Penguin Books.
Sayer, Andrew. 2005. *The Moral Significance of Class*. Cambridge, MA: Cambridge University Press.

Scanlan, John. 2005. *On Garbage*. London, UK: Reaktion Books.
Schiebinger, Londa. 1989. *The Mind Has No Sex? Women in the Origins of Modern Science*. Cambridge, MA: Harvard University Press.
– 1993. *Nature's Body: Gender in the Making of Modern Science*. Boston: Beacon Press.
Schliesmann, Paul. 2011a. "Roots of Mass-Recycling Can Be Traced to Kitchener Man." *Kingston Whig-Standard*, 18 July.
– 2011b. "Reduce, Reuse Revamp?" *Kingston Whig-Standard*, 19 July.
– 2012a. "Special Report: Green Challenge." *Kingston Whig-Standard*, 12 March.
– 2012b. "'Planet Is a Dumping Ground'" *Kingston Whig-Standard*, 21 August.
– 2015. "City Looks to Increase Recycling" *Kingston Whig-Standard*, 15 January.
Schnaiberg, Allan. 1980. *The Environment: From Surplus to Scarcity*. New York: Oxford University Press.
Schnoor, Jerald L. 2012. "Extended Producer Responsibility for E-Waste" *Environmental Science and Technology* 46, no. 15: 7927.
Schrader, Astrid. 2012. "Haunted Measurements: Demonic Work and Time in Experimentation." *Differences* 23, no. 3: 119–60.
Scott, Heidi V. 2008. "Colonialism, Landscape and the Subterranean." *Geography Compass* 2, no. 6: 1,853–69.
Scott, Rebecca. 2012. "Public Perspectives on the Utilization of Human Placentas in Scientific Research and Medicine." Unpublished paper.
Serres, Michel. 2011. *Malfeasance: Appropriation Through Pollution?* Translated by Anne-Marie Feenberg-Dibon. Stanford, CA: Stanford University Press.
Shackleton, Ryan. 2012. "'Not Just Givers of Welfare': The Changing Role of the RCMP in the Baffin Region, 1920–1970." *Northern Review* 36: 5–26.
Shackley, Simon, and Brain Wynne. 1996. "Representing Uncertainty in Global Climate Change Science and Policy: BoundaryOrdering Devices and Authority." *Science, Technology and Human Values* 21, no. 3: 275–302.
Shadian, Jessica. 2006. "Remaking Arctic Governance: The Construction of an Arctic Inuit Policy." *Polar Record* 42, no. 3: 249–59.
Shapiro, Stuart. 1997. "Caught in a Web: The Implications of Ecology for Radical Symmetry in STS." *Social Epistemology* 11: 97–110.
Shaw, Ian G.R. 2012. "Toward an Evental Geography." *Progress in Human Geography* 36, no. 5: 613–27.
Shotyk, William, and Michael Krachler. 2009. "Determination of Trace

Element Concentrations in Natural Freshwaters: How Low Is 'Low,' and How Low Do We Need to Go?" *Journal of Environmental Monitoring* 11, no. 10: 1747–53.

Shotyk, William, Michael Krachler, Bin Chen, and James Zheng. 2005. "Natural Abundance of Sb and Sc in Pristine Groundwaters, Springwater Township, Ontario, Canada, and Implications for Tracing Contamination from Landfill Leachates." *Journal of Environmental Monitoring* 7, no. 12: 1238–44.

Shove, Elizabeth. 2003. *Comfort, Cleanliness and Convenience the Social Organization of Normality*. New York: Berg.

– 2010. "Beyond the ABC: Climate Change Policy and Theories of Social Change." *Environment and Planning A* 42, no. 6: 1273–85.

Shove, Elizabeth, and Gordon Walker. 2007. "CAUTION! Transitions Ahead: Politics, Practice, and Sustainable Transition Management." *Environment and Planning A* 39, no. 4: 763–70.

ShraderFrechette, Kristin. 2000. "Duties to Future Generations, Proxy Consent, Intra and Intergenerational Equity: The Case of Nuclear Waste." *Risk Analysis* 20, no. 6: 771–8.

– 2005. "Mortgaging the Future: Dumping Ethics with Nuclear Waste." *Science and Engineering Ethics* 11, no. 4: 518–20.

Sillitoe, Paul. 2006. *Local Science vs Global Science*. New York: Berghah Press.

Simon, Mary. 2009. "Inuit and the Canadian Arctic: Sovereignty Begins at Home." *Journal of Canadian Studies* 43, no. 2: 250–60.

Simpson, Audra. 2014. *Mohawk Interruptus: Political Life Across the Borders of Settler States*. Durham, NC: Duke University Press.

Sinoski, Kelly. 2010. Province Approves 42-Hectare Expansion to Cache Creek Dump. Vancouver Sun, 6 January.

Skill Karin. 2008. "(Re)Creating Ecological Action Space: Householders' Activities for Sustainable Development in Sweden." PhD thesis, Linköping University, Sweden.

Slack, Jennifer Daryl. 1996. "The Theory and Method of Articulation in Cultural Studies." In *Stuart Hall: Critical Dialogues in Cultural Studies*, edited by David G. Morley and Kuan-Hsing Chen, 113–29. London, UK: Routledge.

Smith, Mick. 2011. *Against Ecological Sovereignty: Ethics, Biopolitics, and Saving the Natural World*. Minneapolis, MN: University of Minnesota Press.

Smith, R.J. 1982. "The Risks of Living near Love Canal." *Science* 217: 808–9, 811.

Smuts, Barbara. 1985. *Sex and Friendship in Baboons.* New Brunswick and London: Aldine Transaction.

Solomon, Barry D., Mats Andrén, and Urban Strandberg. 2009. "Thirty Years of Social Science Research on HighLevel Nuclear Waste: Achievements and Future Challenges." Paper presented at the Conference on Managing Radioactive Waste: Problems and Challenges in a Globalized World, University of Gothenburg, Sweden, 15–17 December.

Sonnenberg, Monte. 2011. "Producers, Consumers Already Pay Fair Share." *Kingston Whig-Standard*, 22 September.

Southcott, Chris. 2012. "Can Resource Development Make Arctic Communities Sustainable? Resources and Sustainable Development in the Arctic." *Northern Public Affairs* (spring): 48–9.

Spaargaren, Gert, Arthur P. Mol, and Frederick Buttel, eds. 2006. *Governing Environmental Flows.* Cambridge, MA: MIT Press.

Spatari, Sabrina, Mike Bertram, Robert B. Gordon, Kristopher Henderson, and Thomas Eldon Graedel. 2005. "Twentieth Century Copper Stocks and Flows in North America: A Dynamic Analysis." *Ecological Economics* 54: 37–51.

Spector, Malcom, and John I. Kitsuse. 1972. *Constructing Social Problems.* Menlo Park, CA: Cummings Publishing.

Spelman, Elizabeth V. 2011. "Combing through Trash: Philosophy Goes Rummaging." *Massachusetts Review* 52, no. 2: 313–25.

Spence, Alexa, and Ellen Townsend. 2006. "Examining Consumer Behavior toward Genetically Modified (GM) Food in Britain." *Risk Analysis* 26, no. 3: 657–70.

Sperling, Tony. 2014. "Iqaluit Fire Control." Presentation given to Iqaluit City Council, 30 June.

Spivak, Gayatri Chakravorty. 1988. "Can the Subaltern Speak?" In *Marxism and the Interpretation of Culture*, edited by Cary Nelson and Lawrence Grossberg, 271–313. Urbana: University of Illinois Press.

Stafford, Tori. 2011. "Kingston Investigates Thermal Treatment for Waste." *Kingston Whig-Standard*, 27 November.

Stang, John. 1998. "Tainted Tumbleweeds Concern Hanford." *Tri-City Herald*, 27 December.

Statistics Canada. 2005. *Human Activity and the Environment: Solid Waste in Canada.* Catalogue no. 16-201-X20050008657. Ottawa, ON: Statistics Canada.

– 2008a. *Waste Management Industry Survey: Business and Government Sectors, 2006.* Catalogue no. 16F0023X2006001. Ottawa, ON: Statistics Canada.

- 2008b. *Human Activity and the Environment: Solid Waste in Canada, 2007 and 2008*. Catalogue no. 16-201-X2007000. Ottawa, ON: Statistics Canada.
- 2010. *Waste Management Industry Survey: Businesses and Government Sectors, 2008*. Catalogue no. 16F0023X2010001. Ottawa, ON: Statistics Canada.
- 2011a. *Households and the Environment Survey (HES)*. Record no. 3881. Ottawa, ON: Statistics Canada.
- 2011b. *Focus on Geography Series, 2011 Census*. Catalogue no. 98-310-XWE2011004. Ottawa, ON: Statistics Canada.
- 2011c. "Table 5: Age distribution and median age of Inuit by area of residence – Inuit Nunangat, Canada, 2011." National Household Survey, 2011. Catalogue no. 99-011-X2011001. Ottawa, ON: Statistics Canada. http://www12.statcan.gc.ca/nhs-enm/2011/as-sa/99-011-x/2011001/tbl/tbl05-eng.cfm.
- 2012a. *Human Activity and the Environment: Waste Management in Canada*. Catalogue no. 16-201-X201200011679. Ottawa, ON: Statistics Canada.
- 2012b. "Selected Income Characteristics of Census Families by Family Type – Median Total Income, by Family Type, by Province and Territory." Table 111-0009-CANSIM. Ottawa, ON: Statistics Canada. https://www150.statcan.gc.ca/t1/tbl1/en/tv.action?pid=1110000901.
- 2013a. "Smoking 2013." Catalogue no. 82-625-X201400114025. Ottawa, ON: Statistics Canada. http://www.statcan.gc.ca/pub/82-625-x/2014001/article/14025-eng.htm.
- 2013b. "NHS Profile – Iqaluit, CY, Nunavut." 2011 National Household Survey. Canada Catalogue no. 99-004-XWE. Ottawa, ON: Statistics Canada. http://www12.statcan.gc.ca/nhs-enm/2011/dp-pd/prof/index.cfm.
- 2017. "Waste Management Industry: Business and Government Sectors, 2014." Ottawa, ON: Statistics Canada. https://www150.statcan.gc.ca/n1/daily-quotidien/170324/dq170324c-eng.htm.

Steinberg, Phillip, and Kimberley Peters. 2015 "Wet Ontologies, Fluid Spaces: Giving Depth to Volume through Oceanic Thinking" *Environment and Planning D: Society and Space* 33, no. 2: 247–64.

Stengers, Isabelle. 1997. *Power and Invention: Situating Science*. Minneapolis: University of Minnesota Press.
- 2000a. "Another Look: Relearning to Laugh." *Hypatia* 15, no. 4: 41–54.
- 2000b. *The Invention of Modern Science*. Minneapolis: University of Minnesota Press.

Stern, Paul C. 1999. "Information, Incentives, and Proenvironmental Consumer Behavior." *Journal of Consumer Policy* 22, no. 4: 461–78.

Stevenson, Lisa. 2012. "The Psychic Life of Biopolitics: Survival, Cooperation, and Inuit Community." *American Ethnologist* 39, no. 3: 592–613.

– 2014. *Life Beside Itself: Imagining Care in the Canadian Arctic*. Berkeley, CA: University of California Press.

Stow, Jason P., Jim Sova, and Ken J. Reimer. 2005. "The Relative Influence of Distant and Local (DEW-line) PCB Sources in the Canadian Arctic." *Science of the Total Environment* 342, no. 1–3: 107–18.

Stang, John. 1998. "Tainted Tumbleweeds Concern Hanford." *Tri-City Herald*, 27 December. https://www.mail-archive.com/nativenews@mlists.net/msg01073.html Accessed 8 September 2015.

Strathern, Marilyn. 2003. "Re-Describing Society." *Minerva* 41, no. 3: 263–76.

Streeper, Charles, Julia Whitworth, and J. Andrew Tomkins. 2009. "Lack of International Consensus on the Disposition and Storage of Disused Sealed Sources." *Progress in Nuclear Energy* 51, no. 2: 258–67.

Sturgis, Patrick, Helen Cooper, and Chris Fife-Schaw. 2005. "Attitudes to Biotechnology: Estimating the Opinions of a Better-Informed Public." *New Genetics and Society* 24, no. 1: 31–56.

Suchman, Lucy. 2005. "Affiliative Objects." *Organization* 12, no. 3: 379–99.

Sun, Monic, and Remi Trudel. 2017. "The Effect of Recycling Versus Trashing on Consumption: Theory and Experimental Evidence." *Journal of Marketing Research* (April): 293–305.

Switzer J. 2008. "Having a Little Enviro-Guilt Can Be a Good Thing." *Kingston Whig-Standard*, 16 July.

Szasz, Andrew. 2007. *Shopping Our Way to Safety: How We Changed from Protecting the Environment to Protecting Ourselves*. Minneapolis: University of Minnesota Press.

Szerszynski, Bronislaw, and John Urry, eds. 2010. "Changing Climates: Introduction." Special double issue of *Theory, Culture and Society* 27, no. 2–3: 1–8.

Tagaq, Tanya. 2014. "Eating Seal Meat Is a Vital Part of Life in My Community." *Vice Munchies*, 23 October. https://www.vice.com/en_us/article/z4gdjy/eating-seal-meat-is-a-vital-part-of-life-in-my-community.

Takai, Yasushi, Osamu Tsutsumi, Yumiko Ikezuki, Hisahiko Hiroi, Yutaka Osuga, Mikio Momoeda, Tetsu Yano, and Yuji Taketani. 2000. "Estrogen Receptor-Mediated Effects of a Xenoestrogen, Bisphenol A,

on Preimplantation Mouse Embryos." *Biochemistry and Biophysical Research Communications* 270, no. 3: 918–21.
Tester, Frank J. 2010a. "Can the Sled Dog Sleep? Postcolonialism, Cultural Transformation and the Consumption of Inuit Culture." *New Proposals: Journal of Marxism and Interdisciplinary Inquiry* 3, no. 3: 7–19.
– 2010b. "Mad Dogs and (Mostly) Englishmen: Colonial Relations, Commodities, and the Fate of Inuit Sled Dogs." *Études/Inuit/Studies* 34, no. 2: 129–47.
Tester, Frank J., and Peter Irniq. 2007. "Inuit Qaujimajatuqangit: Social History, Politics and the Practice of Resistance." *Arctic* 61: 48–61.
Tester, Frank J., and Peter Kulchyski. 1994. *Tammarniit (Mistakes): Inuit Relocation in the Eastern Arctic 1939–63*. Vancouver, BC: University of British Columbia Press.
Thomassin-Lacroix, Eric. 2015. "Site Remediation of the Former DEW Line Site at FOX-3 Dewar Lakes, Nunavut." 2015 RPIC Federal Contaminated Sites Regional Workshop, Edmonton, Alberta.
Thompson, Charles. 1969. "Patterns of Housekeeping in Two Eskimo Settlements." Prepared for Department of Indian Affairs and Northern Development. http://publications.gc.ca/collections/collection_2017/aanc-inac/R42-4-1969-1-eng.pdf.
Thompson, Jeremy, and Honor Anthony. 2008. "The Health Effects of Waste Incinerators: 4th Report of the British Society for Ecological Medicine." Preface to the *Report of the British Society for Ecological Medicine*, 2nd edition. London, UK: British Society for Ecological Medicine.
Thomson, Vivian E. 2009. *Garbage In, Garbage Out: Solving the Problems with Long-Distance Trash Transport*. Charlottesville: University of Virginia Press.
Tibbetts, Janice. 2013. "Garbage Collection Is 'One of the Most Hazardous Jobs.'" *Canadian Medical Association Journal* 185, no. 7: E284.
Timmerman, Peter. 2003. "Ethics of High Level Nuclear Fuel Waste Disposal in Canada: Background Paper." Toronto, ON: Nuclear Waste Management Organization (NWMO).
Todd, Zoe. 2015. "Indigenizing the Anthropocene." In *Art in the Anthropocene: Encounters among Aesthetics, Politics, Environment and Epistemology*, edited by Heather Davis and Etienne Turpin, 241–54. Ann Arbor, MI: Open Humanities Press.
– 2016. "An Indigenous Feminist's Take on the Ontological Turn:

'Ontology' Is Just Another Word for Colonialism." *Journal of Historical Sociology* 29, no. 1: 4–22.

Toomey, Carrie. 2008. "It's Easier Than You Think to Put a Lid on Coffee Cup Waste." *Kingston Whig-Standard*, 17 November.

Townsend, Ellen, David Clarke, and Betsy Travis. 2004. "Effects of Context and Feelings on Perceptions of Genetically Modified Food." *Risk Analysis* 24, no. 5: 1369–84.

Tripp, Rob. 2000. "New Seepage Suspected at Former Belle Park Dump." *Kingston Whig-Standard*, Environmental Bureau of Investigation, 6 October.

Tsing, Anna. 2005. *Friction: An Ethnography of Global Connection*. Princeton, NJ: Princeton University Press.

Tuhiwai Smith, Linda. 2012. *Decolonizing Methodologies: Research and Indigenous Peoples*. Second Edition. London, UK: Zed Books.

UNESCO International School of Science for Peace. 1998. *Nuclear Disarmament, Safe Disposal of Nuclear Materials or New Weapons Developments?: Where Are the National Laboratories Going?* Edited by Paolo CottaRamusino, Giuseppe Gherardi, Antonino Lantieri, Vladimir Kouzminov, Maurizio Martellini and Rosanna Santesso. Venice, Italy: UNESCO Venice Office.

UN-Habitat. 2010. *Solid Waste Mangaement in the World's Cities: Water and Sanitation in the World's Cities*. London, UK: Earthscan. https://unhabitat.org/books/solid-waste-management-in-the-worlds-cities-water-and-sanitation-in-the-worlds-cities-2010-2.

United Nations Environment Programme (UNEP). 2009. "Recycling: From E-Waste to Resources" Sustainable Innovation and Technology Transfer Industrial Sector Studies. Nairobi, Kenya: United Nations Environment Programme. http://www.unep.fr/scp/publications/details.asp?id=DTI/1192/PA.

– 2010. *Metal Stocks in Society: Scientific Synthesis*. International Panel for Sustainable Resource Management, Working Group on the Global Metal Flows. Nairobi, Kenya: United Nations Environment Programme. www.resourcepanel.org/file/387/download?token=XhxT85ju.

– 2011 (1989). *Basel Convention: On the Control of Transboundary Movements of Hazardous Wastes and Their Disposal*. Nairobi, Kenya: United Nations Environment Programme.

United States Department of Energy. 2006. "Recycling Paper and Glass." Washington, DC: US Energy Information Administration.

*Up Here: Life in Canada's Far North Magazine*. 2014. July/August. https://uphere.ca/issues/julyaugust-2014.

Urban Systems. 2002. "Village of Ashcroft Economic Development Strategy." Ashcroft, BC: Village of Ashcroft.

Urry, John. 2002. *Global Complexity*. Oxford, UK: Polity Press.

Usher, Peter J. 2000. "Traditional Ecological Knowledge in Environmental Assessment and Management." *Arctic* 53, no. 2: 183–93.

Utilities Kingston. 2019. *Information Report to Environment, Infrastructure and Transportation Policies Committee*. Report Number EITP-19-003. 6 March.

VandenBrink, Danielle. 2012. "Appeal of One-Bag Policy Falls Flat." *Kingston Whig-Standard*, 16 August.

Van de Poel, Ibo. 2008. "The Bugs Eat the Waste: What Else Is There to Know? – Changing Professional Hegemony in the Design of Sewage Treatment Plants." *Social Studies of Science* 38, no. 4: 605–34.

Van Ewijk, Stijn, and Julia Stegemann. 2014. "Limitations of the Waste Hierarchy for Achieving Absolute Reductions in Material Throughput." *Journal of Cleaner Production* 132: 122–8.

Van Gulck, Jamie, and Richard Dwyer. 2012. "Solid Waste Survey in the Territories." *Journal of the Northern Territories Water and Waste Association* (September 2012): 24–5.

Van Oostdam, J., S. G. Donaldson, M. Feeley, D. Arnold, P. Ayotte, G. Bondy, L. Chan, et al. 2005. "Human Health Implications of Environmental Contaminants in Arctic Canada: A Review." *Science and the Total Environment*, 351–2: 165–246.

van Vliet, Bas, Heather Chappells, and Elizabeth Shove. 2005. *Infrastructures of Consumption: Environmental Innovation in the Utility Industries*. London, UK: Earthscan.

van Wyck, Peter C. 1997. *Primitives in the Wilderness: Deep Ecology and the Missing Human Subject*. Albany: State University of New York Press.

– 2002a. "The American Monument." In *Alphabet City: Lost in the Archives*, no. 8, edited by Rebecca Comay, 740–67. Cambridge, MA: MIT Press.

– 2002b. "The Highway of the Atom: Recollections Along a Route" *Topia: Canadian Journal of Cultural Studies* 7: 99–115.

– 2004. "American Monument: The Waste Isolation Pilot Plant." In *Atomic Culture: How We Learned to Stop Worrying and Love the Bomb*, edited by Scott C. Zeman and Michael A. Amundson, 149–72. Boulder: University Press of Colorado.

– 2005. *Signs of Danger: Waste, Trauma, and Nuclear Threat*. Minneapolis: Theory Out of Bounds Series. Minneapolis: University of Minnesota Press.

- 2008. "An Emphatic Geography: Notes on the Ethical Itinerary of Landscape." *Canadian Journal of Communication* 33, no. 2: 171–91.
- 2010. *The Highway of the Atom*. Montreal and Kingston: McGill-Queen's University Press.
- 2012. "Northern War Stories: The Dene, the Archive, and Canada's Atomic Modernity." In *Bearing Witness: Perspectives on War and Peace from the Arts and Humanities*, edited by Sherrill Grace, Patrick Imbert, and Tiffany Johnstone, 175–85. Montreal and Kingston: McGill-Queen's University Press.
- 2013a. "An Archive of Threat." *Future Anterior* 9, no. 2: 53–80.
- 2013b. "Innis and I on the Highway of the Atom." In *Harold Innis in the North: Appraisals and Contestations*, edited by William Buxton, 326–55. Montreal and Kingston: McGill-Queen's University Press.
- 2013c. "Footbridge at Atwater: A Chorographic Inventory of Effects." In *Thinking with Water*, edited by Cecilia Chen, Janine MacLeod, and Astrida Neimanis, 256–73. Montreal and Kingston: McGill-Queen's University Press.
- 2014. "Theory in a Cold Climate." *Topia: Canadian Journal of Cultural Studies* 32 (fall): 7–19.

Varga, Peter. 2014a. "City Can't Douse Iqaluit's Latest Massive Dump Fire." *Nunatsiaq News*, 21 May.
- 2014b. "Iqaluit Dump Fire Smoke Not a Public Health Emergency, GN Says." *Nunatsiaq News*, 4 August.
- 2014c. "City of Iqaluit Will Spend $3.3 Million of Its Own Money to Douse Dumpcano." *Nunatsiaq News*, 4 August.
- 2014d. "No End in Sight for Iqaluit Dump Fire, Officials Say." *Nunatsiaq News*, 26 May.
- 2014e. "Hired Hands Dig into Nunavut's Biggest Dump Fire." *Nunatsiaq News*, 2 September.
- 2014f. "Environment Canada: Iqaluit Dump Smoke Tests Reveal Little Immediate Danger." *Nunatsiaq News*, 11 June.
- 2014g. "City Council Orders Iqaluit Fire Department to Extinguish Dump Fire." *Nunatsiaq News*, 12 June.
- 2015. "Iqaluit Sled Dog Owners Lose Winter Lot to Asphalt Plant." *Nunatsiaq News*, 9 January.

Velderman, B.J., R. ZapfGilje, F. Fortin, and D. DuBois. 1997. "Brownfields: A Canadian Perspective." *Remediation Journal* 8: 35–44.

Vicente, Paula, and Elizabeth Reis. 2008. "Factors Influencing Households' Participation in Recycling." *Waste Management and Research* 26, no. 2: 140–6.

Villa-Vicencio, Charles. 2003. "Restorative Justice: Ambiguities and Limitations of a Theory." In *The Provocations of Amnesty: Memory, Justice and Impunity*, edited by Charles Villa-Vicencio and Eric Doxtader, 30–50. Cape Town, South Africa: Africa World Press.

Volk, Tyler. 2004. "Gaia Is Life in a Wasteland of By-Products." In *Scientists Debate Gaia: The Next Century*, edited by Stephen H. Schneider, James R. Miller, Eileen Crist, and Penelope J. Boston, 27–36. Cambridge, MA: MIT Press.

Voosen, Paul. 2012. "Geologists Drive Golden Spike Toward Anthropocene's Base." *E&E News Greenside*, 17 September. http://www.eenews.net/stories/1059970036.

Wachowich, Nancy, in collaboration with Apphia Agalakti Awa, Rhoda Kaukjak Katsak, and Sandra Pikujak Katsak. 1999. *Saqiyuq: Stories from the Lives of Three Inuit Women*. Montreal and Kingston: McGill-Queen's University Press.

Waldby, Catherine, and Robert Mitchell. 2006. *Tissue Economies: Blood, Organs, and Cell Lines in Late Capitalism*. Durham and London: Duke University Press.

Walker, Kenny, and Lynda Walsh. 2012. "'No One Yet Knows What the Ultimate Consequences May Be': How Rachel Carson Transformed Scientific Uncertainty into a Site for Public Participation in Silent Spring." *Journal of Business and Technical Communication* 26: 3–34.

Walker, Mike, Sigfus Johnsen, Sue Rasmussen, Trevor Popp, Jorgen-Peter Steffensen, et al. 2009. "Formal Definition and Dating of the GSSP (Global Stratotype Section and Point) for the Base of the Holocene Using the Greenland NGRIP Ice Core, and Selected Auxiliary Records." *Journal of Quaternary Science* 24, no. 1: 3–17.

Wallsten, Björn. 2013. "*Underneath Norrköping: An Urban Mine of Hibernating Infrastructure.*" PhD diss., Linköping University.

– 2014. "Revenge of the Urks: On Processes of Material Exclusion in Infrastructural Assemblages." Unpublished paper.

Wang, Lizhong, Liping Fang, and Keith W. Hipel. 2011. "Negotiations over Costs and Benefits in Brownfield Redevelopment." *Group Decision Negotiation* 20, no. 4: 509–24.

Waste Diversion Ontario. 2012. "Municipal Data Call: 2012 Diversion Rate by Municipality." Toronto, ON: Waste Diversion Ontario.

Watson, Gavan P.L. 2013. "See Gull: Cultural Blind Spots and the Disappearance of the Ring-Billed Gull in Toronto." In *Trash Animals: How We Live with Nature's Filthy, Feral, Invasive, and Unwanted*

*Species*, edited by Kelsey Nagy and Phillip David Johnson II, 31–8. Minneapolis, MN: University of Minnesota Press.

Watt-Cloutier, Sheila. 2015. *The Right to Be Cold*. Minneapolis, MN: University of Minnesota Press.

Watts, Vanessa. 2013. "Indigenous Place-Thought and Agency amongst Humans and Non-Humans (First Woman and Sky Woman Go on a European World Tour!)." *Decolonization: Indigeneity, Education and Society* 2, no. 1: 20–34.

Wei, Lin, and Yangsheng Liu. 2012. "Present Status of E-Waste Disposal and Recycling in China." *Procedia Environmental Sciences* 16: 506–14.

Whatmore, Sarah J. 2009. "Mapping Knowledge Controversies: Science, Democracy and the Redistribution of Expertise." *Progress in Human Geography* 33, no. 5: 587–98.

Whatmore, Sarah J., and Catharina Landström. 2011. "Flood Apprentices: An Exercise in Making Things Public." *Economy and Society* 40, no. 4: 582–610.

Wheeler, Kathryn and Miriam Glucksmann. 2015. "'It's Kind of Saving Them a Job Isn't It?': The Consumption Work of Household Recycling." *Sociological Review* 63, no. 3: 551–69.

Whitmarsh, Lorraine. 2009. "What's in a Name? Commonalities and Differences in Public Understanding of 'Climate Change' and 'Global Warming,'" *Public Understanding of Science* 18: 401–20.

Widger, William R., Georgiy Golovko, Antonio F. Martinez, Efren V. Ballesteros, Jesse J. Howard, Zhenkang Xu, Utpal Pandya, et al. 2011. "Longitudinal Metagenomic Analysis of the Water and Soil from Gulf of Mexico Beaches Affected by the Deep Water Horizon Oil Spill." *Nature Precedings* (28 February).

Wilk, Richard. 2015. "Afterword: The Waste that Matters." In *Waste Management and Sustainable Consumption: Reflections on Consumer Waste*, edited by Karin M. Ekström, 225–39. New York: Routledge.

Wilkes, James. 2011. "Decolonizing Environmental 'Management': A Case Study of Kitchenuhmaykoosib Inninuwug." MA thesis, Trent University.

Williamson Bathory, Laakkuluk. 2013. "Naamaleqaaq! Idle No More in the Arctic." *Northern Public Affairs* (spring): 39–41.

Wilman, Mary. 2014. "Re: Request for Assistance from Canadian Forces." Email to Ed Zebedee.

Wilson, Catherine. 2004. *Moral Animals: Ideals and Constraints in Moral Theory*. Oxford, UK: Clarendon Press.

Windeyer, Chris. 2007 "Councilors Have Good Laugh Over PETA Request." *Nunatsiaq News*, 27 April.

Wittgenstein, Ludwig. 1994. *The Wittgenstein Reader*, edited by Anthony Kenny. Oxford, UK: Blackwell.

Witze, Alexandra. 2016. "Algae are Melting Away the Greenland Ice Sheet." *Nature* 535, no. 7612 (21 July): 336.

Wolfe, Cary. 2003. "In the Shadow of Wittgenstein's Lion: Language, Ethics and the Question of the Animal." In *Zoontologies: The Question of the Animal*, edited by Cary Wolfe, 1–58. Minneapolis: University of Minnesota Press.

Woolgar, Steve. 1991. "Configuring the User: The Case of Usability Trials." In *A Sociology of Monsters: Essays on Power, Technology and Domination*, edited by John Law, 57–99. London, UK: Routledge.

Worden, Peter. 2014. "A 'Dumpcano' In the Canadian Arctic Has Been Burning for Eight Weeks." *Vice News*, 25 July.

World Health Organization. 2020. *Tobacco*. 27 May.

Wynne, Brian. 1987. *Risk Management and Hazardous Waste: Implementation and Dialectics of Credibility*. Heidelberg, Germany: SpringerVerlag.

– 1992. "Misunderstood Misunderstanding: Social Identities and Public Uptake of Science," *Public Understanding of Science* 1: 281–304.

– 1996. "May the Sheep Safely Graze? A Reflexive View of Expert–Lay Knowledge Divide." In *Risk, Environment and Modernity: Towards a New Ecology*, edited by Scott Lash, Bronislaw Szerszynski, and Brian Wynne, 44–83. London, UK: Sage Publications.

– 2003. "Seasick on the Third Wave? Subverting the Hegemony of Propositionalism: Response to Collins and Evans (2002)." *Social Studies of Science* 33, no. 3: 401–17.

– 2005. "Risk as Globalizing 'Democratic' Discourse?: Framing Subjects and Citizens." In *Science and Citizens: Globalization and the Challenge of Engagement*, edited by Melissa Leach, Ian Scoones, and Brian Wynne, 66–82. London, UK: Zed Books.

– 2006. "Public Engagement as a Means of Restoring Public Trust in Science: Hitting the Notes, but Missing the Music?" *Community Genetics* 9, no. 3: 211–20.

– 2007. "Public Participation in Science and Technology: Performing and Obscuring a PoliticalConceptual Category Mistake." *East Asian Science, Technology and Society: An International Journal* 1: 99–110.

Yildiz, Ebru Demirekler, Kahraman Ünlü, and R. Kerry Rowe. 2004.

"Modelling Leachate Quality and Quantity in Municipal Solid Waste Landfills." *Waste Management and Research* 22, no. 2: 78–92.

Yong Jeong, Byung. 2016. "Occupational Injuries and Deaths in Domestic Waste Collecting Process." *Human Factors and Ergonomics in Manufacturing and Service Industries* 26, no. 5: 608–14.

Yoshizawa (Scott), Rebecca. 2014. "Placentations: Agential Realism and the Science of Afterbirths." PhD thesis, Queen's University.

Yusoff, Kathryn. 2012. "Aesthetics of Loss: Biodiversity, Banal Violence and Biotic Subjects." *Transactions of the Institute of British Geographers* 37, no. 4: 578–92.

– 2013. "Insensible Worlds: Postrelational Ethics, Indeterminacy and the (K)Nots of Relating." *Environment and Planning D: Society and Space* 31, no. 2: 208–26.

Zagozewski, Rebecca, Ian Judd-Henrey, Suzie Nilson, and Lalita Bharadwaj. 2011. "Perspectives on Past and Present Waste Disposal Practices: A Community-Based Participatory Research Project in Three Saskatchewan First Nations Communities." *Environmental Health Insights* 5: 9–20.

Zahara, Alexander. 2015. "The Governance of Waste in Iqaluit, Nunavut." MES thesis, Queen's University.

– 2018. "On Sovereignty, Deficits and Dump Fires: Risk Governance in an Arctic 'Dumpcano.'" In *Inevitably Toxic: Historical Perspectives on Contamination, Exposure, and Expertise*, edited by Brinda Sarathy, Vivien Hamilton, and Janet Brodie, 259–83. Pittsburgh, PA: University of Pittsburgh Press.

Zalasiewicz, Jan. 2008. *The Earth After Us: What Legacy Will Humans Leave in the Rocks?* Oxford, UK: Oxford University Press.

Zalasiewicz, Jan, and Mark Williams. 2015. "First Atomic Bomb Test May Mark the Beginning of the Anthropocene." *The Conversation*, 30 January.

Zalasiewicz, Jan, Colin Waters, Mark Williams, David Aldridge, and Ian Wilkinson. 2017. "The Stratigraphical Signature of the Anthropocene in England and Its Wider Context." *Proceedings of the Geologists' Association*, 610: 1–10.

Zehr, Stephen C. 1999. "Scientists' Representations of Uncertainty." In *Communicating Uncertainty: Media Coverage of New and Controversial Science*, edited by Sharon M. Friedman, Sharon Dunwoody, and Carol L. Rogers, 3–22. Mahwah, NJ: Erlbaum.

# Index

Page numbers in *italics* refer to figures.

AANDC. *See* Department of Indigenous and Northern Affairs (previous incarnations)
actor–network theory (Latour), 154, 159, 174–5
agential realism, 171–3, 176–7, 183, 184
Aglukkaq, Leona, 189, 242n3
air quality monitoring, 192, 197, 202, 210–11
Alaska, US, 115, 118
Alberta tar sands, 231
Alexander, Catherine, 12
Ali, Syed Harris, 20, 36, 230
Alisuag, Tita Bughao, 4–5
Alivaktuk, Julie, *188*
Alma Mater Society (Queen's University), 50–1
Alternatives North, 6
American Can Company, 12
American Chemistry Council, 65
Analytical Services Unit, Queen's University, 25
Anderson, Kay, 129
Anderson, Warwick, 93, 95, 129
*Angry Inuk* (film), 131
"Animal Bodies, Colonial Subjects: (Re)locating Animality in Decolonial Thought" (Belcourt), 26
animals, 126–45; inhuman settler colonialism, 142–5; photographs of, *133, 141*; trash animals, 28–9, 126, 128–32, 134, 144, 145; *tulugaq*, 132–7, *143*; and Western culture, 127–28. *See also qimmiit/qimmiiq* (sled dogs)
*anirniq* (breath or spirits), 130
Anthropocene, 166–85; about, 106, 166, 186; and animals, 129; consequences of, 188–9; as current epoch, 145; and de-stratification, 164, 180; metaphysical theories, 173–6; and species proliferation, 151; waste as signature of, 29, 31, 89, 149, 167–9, 215. *See also* bacteria
Arctic, 101–2, 107, 111–25, 236n1, 239n30. *See also* Distant Early Warning (DEW) Line; Inuit; Iqaluit, NU; military, Canadian; Nunavut; *qimmiit/qimmiiq* (sled

dogs); sovereignty; waste, military; West 40, Iqaluit, NU

Areva (mining company), 229

Arnaquq-Baril, Alethea, 131

arsenic, 6, 10, 115–16, 168, 228, 241–2n1

asbestos, 119

Asch, Michael, 101

Asset Recovery plant, North Bay, ON, 73–4

assimilation and self-determination. *See under* Inuit

Australia, 26, 205, 227

Baarschers, William, 67

BachTech mining corporation, 7–8

bacteria: in the dark biosphere, 185; dealing with waste, 144, 152–3, 156–8, 170, 173, 181; de-stratification, 159–60, 163, 167; Hird on, 22, 30, 149; and increase in biodiversity, 151–3, 162–4, 169, 171; microbes/microorganisms, 30, 151–2, 155–7, 168–9, 240n1; and movement of strata, 30–1, 178; and the West 40 fire, 122. *See also* re-stratification

Baffin Island, 237–8n18, 239n7

bag limits, 42, 47, 49–51, 73

Barad, Karen, 171–3, 177–8, 179–80, 183

*Basel Convention on the Control of Transboundary Movements of Hazardous Wastes and Their Disposal*, 3–4, 229

Baudrillard, Jean, 12

Baviskar, Amita, 144

Beck, Ulrich, 36, 206

beef, contaminated, 170

Belcourt, Billy-Ray, 26, 143

Belkorp (landfill owners), 8

Belle Park landfill, Kingston, ON, 44

beluga, 132

Benoit, France, 228

berries, 210–12

"Beyond the Blue Box: Ontario's Fresh Start on Waste Diversion and the Circular Economy" (Environmental Commissioner of Ontario), 224

BFI (WM industry), 17

biocides, 162–3

biological diversity, 149–54. *See also under* bacteria

biosolids, 235n8

birds, migratory, 132–4

bisphenol A (BPA), 179

black-boxing (Latour), 175–6, 242n3

blue box program, 47, 78, 224–5, 235

Brassier, Ray, 173–4

Brechin, ON, 81

British Petroleum (BP), 109, 155–7

Bulkeley, Harriet, 60, 61–2, 63, 67, 76, 83

Burning Mountain fire, Australia, 205

Cache Creek Landfill (Belkorp), 8–9

Callon, Michel, 36, 206

Cambodia, 10, 229

Cameron, Emilie, 130

Canada: 1990 Green Plan and the Arctic Environmental Strategy, 108; Department of Environment

Canada, 3–4, 16, 189, 192, 193, 209, 242n3; Department of Foreign Affairs, 4; Department of Health Canada, 189, 192, 196, 209, 242–3nn3–4; Department of National Defence (DND), 117, 121–2, 192; differing standards of acceptable levels, 202, 206, 208, 212–13; E-numbers to Inuit, 96, 237n15; Environmental Protection Act, 235n8; export of waste, 9, 42, 69, 81, 229–30, 233n2; Green New Deal, 222; "Northern Strategy," 102; racialized paternal governance in the North, 94–6; treaty negotiations, 26. *See also* Arctic; Department of Indigenous and Northern Affairs; government; military, Canadian; waste, military

Canadian Arctic Resources Committee, 119

Canadian Environmental Protection Act, 20–1

Canadian military. *See* military, Canadian; waste, military

Canadian Union of Public Employees (CUPE), 82

"Can Mother Nature Take a Punch? Microbes and the BP Oil Spill in the Gulf of Mexico" (Hazen), 156–7

Canol pipeline, 115

Cape Dyer, NU, *120*, *121*. *See also* Distant Early Warning (DEW) Line

Cape Romanzof, US, 123

Cardenas, Miriam, 135

Carson, Rachel, 219–20

Cataraqui River, ON, 44

Cavell, Janice, 97

Chakrabarty, Dipesh, 27, 186–7

Chernobyl nuclear disaster, 195

China, 109

Chronic Incorporated, 3, 4–5

"Circumpolar Inuit Declaration on Resource Development Principles in Inuit Nunaat, A" (ICC), 107

"Circumpolar Inuit Declaration on Sovereignty in the Arctic, A" (ICC), 107

Clark, Nigel, 30, 149, 157, 226

Clarke Belt, 167

climate change, 7, 81, 101, 107, 195

"Combing through Trash: Philosophy Goes Rummaging" (Spelman), 13

Community Monitoring Committee (CMC), Simcoe County, 52

compost, 43–4, 73, 235n8

Conference Board of Canada, 10

construction, renovation, and demolition (CRD) waste, 11, 15

Container Corporation of America, 65

contaminated sites, cleanup of: financial responsibility for, 6, 114, 115–16, 228; Love Canal, 234n1; principles of, 107, 239n30; research on, 27. *See also* Canada; Distant Early Warning (DEW) Line; landfills; open dumps; United States (US); waste, military; West 40, Iqaluit, NU

contaminant levels, 35, 193, 206, 208, 243n4, 243n5. *See also* Distant Early Warning (DEW) Line; West 40, Iqaluit, NU

Cooper, Helen, 194
Cornwall, ON, 117
cosmologies: Indigenous/Inuit, 28–9, 127, 130, 132, 227, 240n1, 241n8; Western, 28–9, 106, 110, 144
Cosmos 954 (Soviet satellite), 115, 167
Coulthard, Glen, 26
Cournoyea, Nellie, 106
Cronon, William, 236n4
Crutzen, Paul, 166
cutting-together-apart (Barad), 172–6, 177, 181–5

dark biosphere, 185
Davis, Mike, 20
Deepwater Horizon oil spill, 2010, 155–7
*Defence Environmental Strategy* (Department of National Defence), 117
DeGeneres, Ellen, 131
De Landa, Manuel, 165
Deleuze, Gilles, 154–5, 163, 182
Dell, Reconnect program, 15
DeLorenzo, Amy, 230
Dené First Nation, 6, 26, 116, 228
Denmark, 237–8n18
Department of Environment Canada, 3–4, 16, 189, 192, 193, 209, 242n3
Department of Foreign Affairs, 4
Department of Health Canada, 189, 192, 196, 209, 242–3nn3–4
Department of Indigenous and Northern Affairs (previous incarnations): Department of Aboriginal Affairs and Northern Development Canada (AANDC), 101, 121–2, 192; Department of Indian Affairs and Northern Development, 6, 115–16, 119; Department of Indian and Northern Affairs (DIAND), 114; Department of Northern Affairs and Natural Resources, 94, 98, 237n12
Department of National Defence (DND), 117, 121–2, 192
Derrida, Jacques, 160
de-stratification, 154–5, 159–60, 162–5, 167, 180, 182–4
De Vos, Rick, 132, 241n13
de Vries, Gerard, 39
Dewey, John, 38
dioxin levels, 193, 199–203, 207–10
Distant Early Warning (DEW) Line, 117–25; about, 92, 97–8, 118–19, 129; and Arctic sovereignty, 100, 107–8; Cape Dyer, NU, *120, 121*; photographs of, *113, 120, 121*; polychlorinated biphenyls (PCBS) at, 20–1, 119, 122, 123, 125; remediation projects, 25, 27–8, 111, 114, 121–4; Upper Base, Iqaluit, NU, *113*, 114
diversion/diversion rates, 41–2, 47, 51, 72, 83, 234n2. *See also* recycling
Dobbin, Terry, 114
dogs. See *qimmiit/qimmiiq* (sled dogs)
Domhoff, G. William, 118
Douglas, Mary, 13
Downs, Anthony, 38
downstream responses. *See* government; landfills; neoliberal

capitalism; recycling; waste, settler-colonial
Ducharme, Heather, 124
Dump Fire Working Group, 192

Earth Day, 65
Easterling, Keller, 160
eco-efficiency mode of governance, 83
*Economic Study of the Canadian Plastic Industry, Market and Waste* (Environment and Climate Change Canada), 16
EfW (energy-from-waste) facility, 5, 8–9, 21–2, 43, 55, 75, 219. *See also* energy, waste from; fly ash
Eldorado Gold Mines, 116
electronics waste, 15, 65–6, 74, 117, 222, 230
Ellesmere Island, 237–8n18
El Sobrante landfill, 135
energy, waste from, 5, 8, 9, 55, 68–9, 219, 234n1. *See also* EfW (energy-from-waste) facility
Eno, Robert, 122
environmental impacts, 42
*Environmental Indicators Report* (Environment and Climate Change Canada), 16
Environmental Protection Act, Canada, 235n8
Environmental Protection Agency (EPA), United States, 162–3
Environmental Sciences Group (ESG), 124
Environment Canada. *See* Department of Environment Canada
European Commission, 59

European Judeo-Christian perspective, 90–1, 236n4
European Union, 131
e-waste, 14–15, 222
exploration rights, 109
extinction, 150, 153–4

falcons, 135
Farish, Matthew, 94, 98, 107–8
FBM (frozen-block method), 6–7, 168
feed the beast, 37, 234n1. *See also* EfW (energy-from-waste) facility
feminist materialism, 30–1
Fife-Schaw, Chris, 194
Finland, 238n23
fly ash, 34–5, 38, 124, 170, 181, 219
food, 170, 240n3
Foucault, Michel, 33, 40, 53, 183
foxes, 123
fracking, 162
Frankfurt School, 18
free bag policy, 42, 47, 49–51, 73
Freud, Sigmund, 13
Frobisher Bay. *See* Iqaluit, NU
Frobisher Bay (body of water), 114, 122, 189
frozen-block method (FBM), 6–7, 168
furan, 193, 199, 203, 207–10

garbology, 14
Gates, Bill, 216
*General Drury* (ship), 117
geoengineering architecture, 135, 240n5
Ghosn, Rania, 224–5
Giant Mine, NT, 5–6, 10, 115, 168, 228, 241–2n1

Gille, Zsuzsa, 60
Gilmartin, David, 144
Global boundary Stratotype Section and Point (GSSP), 166, 186
global South, 229
Glucksmann, Miriam, 58–9
Gordillo, Gastón, 160
government: focus on individualism, 16–17, 37, 42–4, 46–7, 50, 61, 231; public trust of, 36; relationship with waste management industries, 62, 100, 216–17, 223–5, 231, 235n9. *See also* Canada
governmentality (Foucault), 33, 40, 51, 53
Grant, Iain Hamilton, 173–4
Great Pacific Garbage Patch, 116
green bin services, 41, 70, 73–4, 78, 235
greenhouse gas emissions (GHG), 79, 80, 81, 152
Greenland, 27, 118, 151, 237–8n18, 238n23
Green New Deal, Canada, 222
Greenpeace, 131
Gregory, Derek, 109
grey water, 116
GSSP (Global boundary Stratotype Section and Point), 166, 186
*Guardians of Eternity* (film), 228
Guattari, Félix, 154–5, 163, 182
Guelph, Ontario, 36
gulls, 28, 126, 132, 134–6, 142

Halifax, NS, 117
Haraway, Donna, 23, 128
Harman, Graham, 165, 173–6, 179–80, 184

Harper, Stephen, 100–1, 197, 241n12
hawks, 28, *133*, 135
Hazen, Terry, 156–7
HDR (waste management company), 47–8, 56
Health Canada. *See* Department of Health Canada
heavy-duty diesel vehicles (HDDV), 80
Herrera, Jorge, 135
Hetherington, Kevin, 185
*Highway of the Atom* (van Wyck), 22, 116, 227
Hird, Myra: about, 26, 30; on bacteria/inhuman entities, 22, 28, 30, 149; in Iqaluit, NU, 25–6, 28–9, 87–8, 191, 225–6; "Making Waste Management Public (or Falling Back to Sleep)" (Hird, Lougheed, Rowe, and Kuyvenhoven), 24; on ravens, 127; research, 21–3, 111. *See also* Kingston, ON
Hobson, Kersty, 56
Hooper Bay, 123
Hossay, Patrick, 225–6
household hazardous and special wastes (HHSW), 11, 37, 38
Hudson, Ray, 60, 63, 67, 76, 83
Hudson's Bay Company, 92, 99, 236n7
Huebert, Rob, 97, 99, 109
*Human Activity and the Environment* (Statistics Canada), 15
Hungary, 60
hygiene, 126, 129

ICC (Inuit Circumpolar Council), 237n17
Iceland, 27, 118, 238n23
ICI (industrial, commercial, and institutional waste), 11, 45, 204
*iglu*, 92
indeterminacy. *See* waste, indeterminacy
India, 109
Indigenous Peoples: and ongoing settler colonialism, 226; terminology, 92, 236–7n9, 237n10; Treaty 9, 228–9. *See also* Dené First Nation; Inuit; waste, military; waste, settler-colonial
industrial, commercial, and institutional (ICI) waste, 11, 45, 204
inhuman entities: about, 44–5, 173, 188, 240n1; and bacteria, 159; Hird on, 22, 28, 30, 149; and humans, 34, 38–9, 154, 229; Inuit *versus* RCMP, 26, 127–8, 227, 240n1; and metaphysical theories, 173, 176, 182–3; settler colonialism, 142–5; and waste, 137, 184, 223. *See also* trash animals
integrated waste management (IWM), 42, 45–8, 72, 76, 78–9, 140
International Commission on Stratigraphy (ICS), 166
Inuit: assimilation and self-determination, 27–8, 94–6, 101–7, 129, 140, 187, 196; *versus* Canadian government policies, 104–6; hunting, 131, 132; naming customs, 96, 139, 237n15; pre-colonization, 89, 92, 236n3, 238–9n29; shift to settlements, 92, 94–6, 99, 138; terminology of, 96, 237n16. *See also* Arctic; Iqaluit, NU; Nunavut
Inuit Circumpolar Council (ICC), 237n17
Inuit Nunaat, 236n1
Iqaluit, NU, 89, 92–3, 192–3, 236nn7–8. *See also* Arctic; Inuit; Nunavut; waste, military; West 40, Iqaluit, NU
*Iqalummiut*, 104, 236n5

Jacques Whitford (consulting company), 42, 45–7
Japan, 10–11, 116, 227–8
Jazairy, El Hadi, 224–5
Jenness, Diamond, 91
Johansson, Nils, 31, 162, 184
Johnson, Genevieve, 128
Judd, Alexander, 12

Kant, Immanuel, 174
KARC (Kingston Area Recycling Centre), 41, 46, 70–1, 73–4, 75
Kärrholm, Mattias, 181
Keeling, Arn, 230
Kennedy, Greg, 3, 136–7, 171
King, Wendell, 118
Kingston, ON, 40–53; about waste, 24, 41, 49, 53–4, 73–5, 81; about waste management (WM), 70–1, 234n3, 234–5n5; Belle Park landfill, 44; integrated waste management (IWM) study, 72–3, 76
Kingston Area Recycling Centre (KARC), 41, 46, 70–1, 73–4, 75
Kingston City Council, 41

Kingston Rental Property Owners Group, 50–1
*Kingston Whig-Standard*, 49
Kitchenuhmaykoosib Inninuwug (KI), 228–9
knowledge translation, 194–6
Koerth-Baker, Maggie, 205
Korea, 73–4, 81
Krupar, Shiloh, 100, 221, 231
Kuyvenhoven, Cassandra, 24, 25, 40, 59

Lackenbauer, P. Whitney, 94, 98, 106–8, 236n1
La Fay, Howard, 118–19
Laflèche Environmental, 235n6
land, ownership of, 26, 97, 237n17, 238nn19–23. *See also* Inuit; waste, settler-colonial
land, undeveloped, 90–1
landfills: beautification of, 58, 234n1; Belle Park landfill, Kingston, ON, 44; Cache Creek Landfill (Belkorp), 8–9; capacity of, 63, 204–5; design/models of, 178–9, 218; El Sobrante landfill, 135; heterogeneous mix, 30, 153, 158, 162, 170–1, 181–2, 204; lifespan of, 10, 19, 31; mining of, 162; photographs of, 133; plastics in the, 16, 159, 170, 204–5, 213; producer responsibilities, 223; runoff, 114; Site 41, landfill proposal, 52–3, 55; in southern Canada, 20; Vancouver Landfill (City of Vancouver), 8–9. *See also* bacteria; leachate; open dumps; trash animals
Lathers, Marie, 93, 129, 137

Latour, Bruno: on actor-network theories, 154, 164, 174–5; on black-boxing, 242n3; on microbes, 30; on public groups, 24, 34. *See also* Politics-5 (Latour)
leachate: contamination by, 29, 36–7, 153, 158–9, 181, 213; and fires, 189–90, 205, 212, 213; in food chain, 135; heterogeneity of, 170–1, 181–2; in Kingston, ON, 44; and landfill mining, 162–3; in Simcoe County, ON, 52
Leonard, Annie, 235n7
Lepawsky, Josh, 15, 74, 222, 223, 230
Liboiron, Max, 16, 17, 65, 230
Lippmann, Walter, 36–7
Locke, John, 90
Lougheed, Scott, 24, 40, 52, 230
Love Canal, NY, US, 58, 234n1
Lovett, Robert A., 118

"Making Waste Management Public (or Falling Back to Sleep)" (Hird, Lougheed, Rowe, and Kuyvenhoven), 24
Makris, Jim, 5
Malaysia, 10, 229
Manhattan Project (US), 116, 227–8
Maniates, Michael, 63
Margulis, Lynn, 152
Marres, Noortje, 24, 33, 35, 51
Massachusetts Department of Environmental Protection report, 134
Massachusetts Institute of Technology (MIT), 118
McCallum, John, 117–18

McCluskey, Kerry, 26, 136
McGregor, Deborah, 226
mechanical separation solution, 56
Meillassoux, Quentin, 173–4, 180
metaphysical theories, 173–4, 177
Metuzals, Jessica, 24
Metzger, Jonathan, 184
Mexico, 15
mice, 126
microbes/microorganisms, 30, 151–2, 155–7, 168–9, 240n1. *See also* bacteria
Microsoft, 216
military, Canadian: attempted assimilation of Inuit, 94; base at Iqaliut, 28; Operation NANOOK, 100, 104, 197; Operation NUNALIVUT, 100; presence in the North, 104, 106, 108–9, 143, 188. *See also* Distant Early Warning (DEW) Line; sovereignty; waste, military
Milloy, John S., 237n13
minerals, 66
mines/mining companies: Areva (mining company), 229; BachTech mining corporation, 7–8; Eldorado Gold Mines, 116; Giant Mine, NT, 5–6, 10, 115, 168, 228, 241–2n1; Miramar Mining, 6; orphaned, 115; Platinex (mining company), 228–9; Royal Oak Mines, 6
Ministry of the Environment, ON, 44, 52, 63, 234n2
Miramar Mining, 6
missionaries, 89
Mitchell, Scott, 122
Moose Creek, ON, 71, 235n6
moral economy, 58–9

Morrell, Geoff, 157
Munton, Don, 119
Murphy, David, 14
muskox, 138
Myers, Heather, 119

Nagy, Kelsey, 128
*nalunaktuq* (nature's unpredictability), 130
Nansen, Fridtjof, 91
narwhals, 132
Naylor, Thomas, 50
neoliberal capitalism: about, 17; example of, 46–7; and forgetting of waste, 25, 226; modes of governance of waste in, 83; and multinational corporations, 39–40; and parameters for public consultation, 56; and safety and security, 142; and settler colonialism, 127; waste as technological issue, 34. *See also* waste, invisibility/hiding of
Noakes, Jeff, 97
Norterra Organics, 75, 235n8
northern, terminology of, 236n2
North Warning System, 119
Northwest Passage, 109, 237n17
Northwest Territories, 115, 167, 239n7, 241n12
Norway, 237–8n18, 238n23
nuclear waste, 10, 11, 19, 167, 218–19, 243n4
*nuna* (land), 130
Nunatsiavut, 236n1
Nunavik, 236n1
Nunavut, 88, 89, 96, 104, 130, 236n1, 237n14. *See also* Arctic; Inuit; Iqaluit, NU
Nunavut Arctic College, 25–6

Nunavut Land Claims Agreement, 96
Nunavut Research Institute, 25–6
Nunavut Tunngavik, 121–2
Nungak, Zebedee, 96, 236n3

oil industry: Alberta tar sands, 15, 231; development in the Arctic, 101–2, 107; dump sites in the Arctic, 27, 109, 122; focus on extraction, 10, 125, 153, 159, 188; oil prices, 82; spills, 115, 123, 155–8. *See also* government; plastics
Olmer, Naya, 81
Ontario Fisheries Act, 44
open dumps, 20, 28, 61, 93, 136, 153, 204. *See also* landfills; trash animals; West 40, Iqaluit, NU
Operation NANOOK, 100, 104, 197
Operation NUNALIVUT, 100
O'Reilly, Kevin, 119
Ottinger, Gwen, 196
overproduction and overconsumption, 14, 34, 217, 221–2

packaging (and printed paper [PPP]), 11, 46, 64–65, 220
Padlayat, Issacie, 140
Paquiz, Leah, 4
Paris, France, 33
Parizeau, Kate, 223, 230
Paterson, Bryan, 50, 221–2
PCB (polychlorinated biphenyls), 20–2, 119, 121–4, 161, 239n7
Pearl Harbour attack, 115
Pearson, Lester, 238n20
People for the Ethical Treatment of Animals (PETA), 131

*People of the Twilight* (Jenness), 91
Peters, Kimberley, 156–7
Petts, Judith, 56
phenomenology, 174, 177
Philippines, 3–5, 4, 10, 95, 229
pigeons, 126
Pinetree Line, 239n9
pipelines, 102–3, 115, 121
placentas, 11, 233n4–5
plastics: about, 11, 96, 231; bisphenol A (BPA), 179; in DEW Line waste, 11, 96; and dioxins, 201; in the landfills, 16, 159, 170, 204–5, 213; Liboiron on, 230; lobby for, 65; and persistent organic pollutants (POPS), 115–16; recycling of, 18, 42, 50, 73–4; single serving, 13, 224
PlasticsEurope, 74
Pleistocene epoch, 151–2, 167
Point Hope, Alaska, 123
polar bears, 123, 124, 126, 136, 138, 240n2
Politics-5 (Latour), 33–4, 39–40, 42, 44, 52–5
Pollans, Lily B., 67–9, 76
polybrominated diphenyl ether (BPDE), 179
polychlorinated biphenyls (PCB), 20–1, 119, 121–4, 161, 239n7
polystyrene (Styrofoam), 73–4
Pond Inlet, NU, 237–8n18
postcolonialism, 129, 187, 238–9n29
post-consumption waste, 14, 16, 53, 190, 217
poststructuralism, 174
Powell, Maria, 194
power relations, 34, 55, 56

Predko, Hillary, 24
Progressive Waste and Waste Management (company), 61, 70
property taxes, 47, 50, 72, 234n2
public consultation: government requirement, 48; in Kingston, ON, 45–8; late in the process, 55–6; legal requirement, 40, 209; *versus* multinational corporations in decision making, 56; when waste becomes visible, 38–9, 50–1, 221

*Qallunaat*, 92, 103–4, 142, 190, 236n5, 243n2
Qamani'tuaq (Baker Lake), NU, 229
*qammaq*, 92
*Qaujimajatuqangit* (Inuit knowledge), 26, 138. See also Traditional Knowledge (TK)
*Qaujivaallirutissat* (Department of Northern Affairs), 94
Qikiqtani Inuit Association (QIA), 98, 141
Qikiqtani Truth Commission (QTC), 98, 142
*qimmiit/qimmiiq* (sled dogs), 99, 138–42; and colonialism, 143–4, 240–1n7; as domesticated or feral, 138; in Greenland, 241n13; Hird and, 28; killed by RCMP, 127–8, 132, 142, 241n11; photographs of, *141*; roaming of, 139, 141, 241n9
*qimutsiit*, 139
Quebec, 239n7
Queen's University (Alma Mater Society), 50–1

raccoons, 126

Rankin Inlent, 191
Rathje, William L., 14
rats, 126
ravens, 28, 126, 127–8, *141*
RCMP (Royal Canadian Mounted Police), 89, 97, 99, 129, 140–2, 237–8n18
*Reassembling Rubbish: Worlding Electronic Waste* (Lepawsky), 15
recycling, 58–83; blue-box programs in Ontario, 63, 78, 235n10; cost of processing, 64, 225; as false solution, 65, 216–17; incentives to, 43–4, 67, 79; as individual's responsibility, 17; limits of, 38; single stream *versus* two-stream, 47; and Stewardship Ontario, 79; transportation of, 25, 79–80, 229–30, 236n11
Reddy, Chris, 156
Redfern, Madeleine, 137
*Red Skin, White Masks* (Coulthard), 26
relationality, 31, 169, 171–6, 180–2
Remarkable Recyclers program, 44, 57
Renders, Micky, 24, 28
Reno, Joshua, 13, 231
Republic Services, 216
residential school system, 95, 237n13
Resolute Bay, NU, 93
Resolution Island, NU, 122, 124
re-stratification, 31, 159, 160–4, 167–70, 180, 184
Rice, Matt, 130–1
Riha, Jacob, 24
risk assessments, 20, 35–6, 37, 52–3, 55

Rose, Deborah Bird, 227
Rowe, R. Kerry, 22–4, 25, 52
Royal Canadian Mounted Police (RCMP), 89, 97, 99, 129, 140–2, 237–8n18
Rumsfeld, Donald, 19, 35
Russia/Soviet Union, 115, 118, 119, 167, 237–8n18, 238n23
Rutherford, Stephanie, 223–4
Rutter, Allison, 25, 27–8
Ryedale, North Yorkshire, UK, 39

Sahtú Dené, 116, 228
Sanchez, Andrew, 12
Sandlos, John, 230
Sarcpa Lake site, 123
Saul, John Ralston, 237n17
Scanlan, John, 91
Schrader, Astrid, 177
Scott Environmental Services, 234n4, 235n8
seals, 131, 138
*Second Treatise of Government* (Locke), 90
Serres, Michel, 165
sewage systems, 11, 26, 33, 89, 235n8
Shackleton, Ryan, 99
Shadian, Jessica, 102
Shaw, Ian, 175, 176–7
Sherman, Troy, 51
shipping costs, 5
*Silent Spring* (Carson), 219
Simcoe County, ON, 52–4, 234n3
single-serve objects, 12–13
Site 41, landfill proposal, 52–3, 55
Skill, Karin, 59
sled dogs (*qimmiit/qimmiiq*). See *qimmiit/qimmiiq*
Smith, Mick, 229

snow, watermelon, 151
Snow Lake, MB, 7
southern, terminology of, 236n2
sovereignty: Chakrabarty on, 187–8; and colonialism/Canadian government, 100–3, 106, 132, 237–8n18, 238n19–23, 239n1, 241n12; and the DEW Line, 107–8, 118–20, 125; Inuit on, 107, 237n17; "Statement on Canada's Arctic Foreign Policy: Exercising Sovereignty and Promoting Canada's Northern Strategy Abroad," 239n30; and waste issues, 27–8, 97–9, 204, 223; *versus* "wasteland," 91. *See also* Operation NANOOK
Soviet Union/Russia, 115, 118, 119, 167, 237–8n18, 238n23
speculative realism, 173–4, 176–7
Spelman, Elizabeth, 13, 14
Sputnik, 119
Statistics Canada, 15
Steinberg, Phillip, 156–7
Stengers, Isabelle, 164, 183
Stewardship Ontario, 49, 79, 216, 225
St Laurent, Louis, 98
*Story of Stuff* (Leonard), 235n7
strata (Deleuze and Guattari), 31, 154–8
Sturgis, Patrick, 194
Styrofoam (polystyrene), 73–4
Sustainability Kingston, 76
Sverdrup, Otto, 237–8n18
Sweden, 59, 238n23
Szasz, Andrew, 231

Tagaq, Tanya, 131–2
*taima* (stop/enough), *188*, 189

tar sands, 231
technological rationality, 18
*terra nullius*, 101
Tester, Frank J., 236n6, 240-1n7
thermal processing solution, 56
tipping fees, 8-9
Tomlinson Environmental Services, 71, 74-5, 234-5nn4-5
Traditional Knowledge (TK), 26, 98, 127, 195, 226-7. *See also* cosmologies
Train the Trainer, Kingston, ON, 77
Tranter, Martyn, 151
trash animals, 28-9, 126, 128-32, 134, 144, 145. See also *qimmiit/qimmiiq* (sled dogs); *tulugaq* (ravens)
Trash Pack toys, 220
trout, 123
Trudeau, Justin, 5
Trudeau, Pierre, 108
Truth and Reconciliation Commission, 237n13
*tulugaq* (ravens), 132-7, 143
*Tulugaq: An Oral History of Ravens* (McCluskey), 26

uncertainty: about microorganisms response, 157; in government response to West 40 fire, 198, 201, 206-12; for the Inuit, 130, 229; material uncertainty, 32, 191; and waste, 11, 19-20, 218-21
United Kingdom (UK), 59, 117
United States (US): Alaska, 115, 118; Cape Romanzof, 123; disposal of waste at sea, 117; Environmental Protection Agency (EPA), 162-3; e-waste, 222; import of Canada's waste, 9, 42, 73-4; Love Canal, NY, US, 58, 234n1; Manhattan Project, 116, 227-8; Massachusetts Department of Environmental Protection report, 134; Massachusetts Institute of Technology (MIT), 118; militarization in the North, 27, 92, 93, 100, 120-1, 188; nuclear test sites, 108; oil spills, 123; Pearl Harbour attack, 115; Pine Tree Line, 239n8; and sovereignty claims to the Arctic, 237-8n18, 238n23; Washington State, 9; Waste Isolation Pilot Plant, 168. *See also* Distant Early Warning (DEW) Line
"Up and Down with Ecology" (Downs), 38
Upper Base, Iqaluit, NU, *113*, 114. *See also* Distant Early Warning (DEW) Line
upstream issues, 218-30; bypassing of, 25, 37, 46, 52, 58, 77-83; effect upon downstream responses, 222; Kingston, ON, 49, 55-7, 71; as metaphor, 18; military and industrial waste, 62; and settler colonialism, 20. *See also* downstream responses
uranium, 115, 116, 167, 219, 227-8, 229
urks (discards of urban infrastructures), 31, 160-1

Vancouver Landfill (City of Vancouver), 8-9
Vancouver/Metro Vancouver (MV), 3, 8-9, 233n1

Van Vlymen, Diana, 24
van Wyck, Peter, 22, 228, 230
vermin control, 28, 126
voles, 123
Volk, Tyler, 167
von Massow, Michael, 230

walrus, 132
Washington State, US, 9
waste, disposal of: *Basel Convention on the Control of Transboundary Movements of Hazardous Wastes and Their Disposal*, 229; Kingston, ON, 71–2, 77–8; landfills as preferred mode, 61; mode of governance, 61–3, 83; and neoliberal capitalism, 60, 77, 109–10, 111–12, 144; at sea, 26, 116–17, 190, 239n6. *See also* diversion/diversion rates
waste, flow of, 3–32; examples of, 3–9
waste, future of, 215–31; Canada's waste future, 230–1; overproduction and overconsumption, 14, 34, 217, 221–2; relationship of government with industry, 223–5; and social injustice, 225–30; uncertainty, 11, 19–20, 218–21
waste, governance of, 59–69
waste, indeterminacy, 186–215; bacterial creations, 31, 169; and containment, 10, 185; dump fire materiality, 204–5; material politics, 212–14; and metaphysical theories, 177–9; and the public, 54–5; publics, expertise, and trust, 194–6; and strata movement, 180–4; uncertainty of waste, 206–12. *See also* West 40, Iqaluit, NU
waste, infrastructure of, 149–65; biological diversity, 149–54; de-stratification, 154–5, 159–60, 162–5, 167, 180, 182–4; re-stratification, 31, 159, 160–4, 167–70, 180, 184; strata, 154–8. *See also* bacteria; inhuman entities
waste, in Southern Canada, 33–58; assembling politics, 53–5; consultation, 36–8; making waste management public, 51–3; managing risk, 35–6; politics and power, 38–40; politics in the making, 40–51; upstreaming waste issues, 55–7
waste, invisibility/hiding of, 58–83; diverts public, 77–83; in open dumps *versus* landfills, 28; and solutions, 37–8, 44, 48–9, 221; in Southern Canada, 13–14, 33–4
waste, management of: about, 14; diversion, 72–4; government/industry relationship, 61–2, 64, 100, 216–17, 223–5, 231, 235n9; integrated waste management (IWM), 42, 45–8, 72, 76, 78–9, 140; as technological issue, 34–5; waste as resource, 68, 74–6
waste, military, 117–25; at abandoned sites in the North, 26–7, 89, 103, 114, 143, 204; from Americans, 93; amount of, 15–16, 25, 62, 223; Krupar and Reno on, 231; photographs of,

113, 121. *See also* Distant Early Warning (DEW) Line; military, Canadian
waste, municipal solid (MSW): *versus* industrial waste, 64; in Kingston, ON, 41; quantities of, 8, 10–11, 15–16, 233n1; separated from industrial waste, 204; transportation of, 79–80, 236n11
waste, overview: about, 11–14, 90, 236n4; compost, 43–4, 73, 235n8; definition of, 11–12, 233n4; electronics waste, 15, 65–6, 74, 117, 222, 230; export of, to United States, 42, 69, 73–4; hierarchy of, 59, 60, 160; industrial, commercial, and institutional waste (ICI), 11, 45, 204; and Others, 129; photographs of, 46, 62, 88, *113*, *120*, *121*, *188*, 220; post-consumption waste, 14, 16, 53, 190, 217; sewage systems, 11, 26, 33, 89, 235n8; studying Canada's waste issues, 21–32; as symptom of social injustice, 20, 225–30; terminology of, 112, 239n1; transportation of, 3–5, 66–7, 71–2, 75, 81, 223; waste as signature of Anthropocene, 29, 31, 89, 149, 167–9, 215. *See also* downstream responses; landfills; leachate; open dumps; upstream issues
waste, settler-colonial, 87–110; in abandoned military sites, 114–15; Hird on, 26–7; history of, 236n6; and inhumans, 143–5; Inuit self-determination, 103–7; management of Others, 127–8; modelling to Inuit, 89, 92–6; security, 100–3; in southern Canada, 236n2; sovereignty, 97–100; and *terra nullius*, 101; on undeveloped land, 90–1, 109–10; waste as symptom of, 111–12. *See also* waste, military
Wastech Services (a Belkorp subsidiary), 9
Waste Diversion Act, ON, 63
Waste Diversion Ontario (WDO), 225
Waste Free Ontario Act, 225
Waste Isolation Pilot Plant, US, 168
Waste Logix, Brechin, ON, 74
Waste Management Incorporated, 17–18, 41, 70, 71, 234–5n5
water, drinking, 172
Watson, Gavan, 136
Watson, Matt, 60, 63, 67, 76, 83
Watt-Cloutier, Sheila, 213, 237n17
Weaver, Paul, 60
West 40, Iqaluit, NU, 186–214; about, 112–14, 190, 204, 225–6; dioxin levels, 193, 199–203, 207–10; expertise messaging on, 197–204, 207–8; fire, May 2014, 239n4; fires previous to May 2014, 112, 189, 242n1; local residents response, 29, 209–12, 213; open dump *versus* landfill, 213, 239n1, 243n6; photographs of, *188*; ravens at, 127; and settler colonialism, 188, 190, 191–3, 213–14; visibility of, 87, 190–1
Wheeler, Kathryn, 58–9
white privilege, 23, 26. *See also* Inuit; *Qallunaat*
Whitmarsh, Lorraine, 195

Wıìlıìdeh Yellowknives Dené First Nation, 228
wildlife management, 134–5
Wilk, Richard, 11, 13–14
Wilson, J.H., 141–2
wolves, 136
World Health Organization (WHO), 199
Wynne, Brian, 195, 204

XL Foods, 170

Yatsushiro, Toshio, 140
Yellowknife, NWT, 115–16, 132–4
Yukon, 115
Yusoff, Kathryn, 150, 158

Zahara, Alexander: with Hird in Iqaluit, NU, 28–9, 87–8; on Inuit perspectives on waste, 103–4, 127, 197; on waste and settler colonialism, 111–12, 191, 211; mentions, 24, 26, 31–2